# CLOSET SPACE

*Closet Space* examines how the spatial metaphor of 'the closet' works not merely metaphorically to describe the concealment, denial and ignorance of gay men's presence in the world, but also materially as a spatial practice of power/knowledge. It explores how our use of this metaphor draws on implicit assumptions about space and movement to argue that the closet can be found at a variety of spatial scales, from the body to the globe, and in numerous locations.

By situating the closet in different locations and at different spatial scales, Michael P. Brown shows how its iterations have implications for a wide array of contemporary social theories and debates. Drawing on both qualitative and quantitative means, the author explores the closet through texts that include: the oral histories of gay men in the UK and the US, the sexualised landscape of a New Zealand city, the national census of Britain and the US, and international travelogues. These multiple placements speak to/through Butler's theory of performativity, Lefebvre's urban theory, Foucault's notion of governmentality, and Lacan's arguments about desire.

By tacking between metaphoric and material spaces and scales of the closet, and using a variety of research techniques and materials, this book provides a decidedly geographic extension of previous research on the closet, which has largely come from literary criticism and the humanities. While *Closet Space* augments the growing interest in sexuality and space specifically, its broad theoretical and empirical reach should interest scholars working in social theory more generally.

**Michael P. Brown** is Assistant Professor of Geography, University of Washington, Seattle, USA.

# CRITICAL GEOGRAPHIES

Edited by Tracey Skelton

*Lecturer in International Studies, Nottingham Trent University*

and Gill Valentine

*Professor of Geography, The University of Sheffield.*

This series offers cutting-edge research organised into four themes: concepts, scale, transformations and work. It is aimed at upper-level undergraduates, research students and academics, and will facilitate interdisciplinary engagement between geography and other social sciences. It provides a forum for the innovative and vibrant debates which span the broad spectrum of this discipline.

# CLOSET SPACE

Geographies of metaphor from the
body to the globe

*Michael P. Brown*

London and New York

First published 2000
by Routledge
2 Park Square, Milton Park, Abingdon, Oxon, OX14 4RN

Simultaneously published in the USA and Canada
by Routledge
270 Madison Ave, New York NY 10016

*Routledge is an imprint of the Taylor & Francis Group*

Transferred to Digital Printing 2007

Typeset in Perpetua by
RefineCatch Ltd, Bungay, Suffolk

*British Library Cataloguing in Publication Data*
A catalogue record for this book is available from the British Library

*Library of Congress Cataloging in Publication Data*
Brown, Michael P.
Closet space: geographies of metaphor from the body to the globe/
Michael P. Brown.
p. cm.—(Critical geographies)
Includes bibliographical references and index
1. Closeted gays.   2. Coming out (Sexual orientation)   3. Spatial
behaviour.   4. Social sciences—Philosophy.   I. Title.   II. Series.
HQ76.25.B77   2000   00–027301
306.76′6—dc21

ISBN 0–415–18764–8 (hbk)
ISBN 0–415–18765–6 (pbk)

To Jonathan,
who likes to tell his story

# CONTENTS

# FIGURES

# TABLES

# ACKNOWLEDGEMENTS

This book could not have been written without the help of a number of people and institutions. From 'the field' there are several people in the closet whom I thank for sensitising me to its manifold and multiple exercises of power. The powers of the closet prevents me from naming them through my respect for their privacy. They are people very close and very far away from me, but to all of them I say thanks for imbuing this very public piece of scholarship with their very personal and private experiences. More publicly, big thanks also must go to Anna, Beverly, Sasha, Kate, Jenine, David, Peter and the New Zealand Prostitutes' Collective for their interest, encouragement and support during my time in New Zealand. I am grateful also to Neil Miller for our discussions and our travels across Cambridge and Somerville for the perfect cup of coffee.

I am very lucky to work in a discipline inhabited by interested, caring and supportive colleagues. Their compliments and their criticisms have helped me so much. Thanks to: John Adams, David Atkinson, Paul Boyle, Bruce Braun, Jenny Cameron, David Conradson, Michael Dear, Robyn Dowling, Jim Duncan, Dave Gibbs, Kathy Gibson, Jamie Gosch, Derek Gregory, Matt Hannah, J.W. Harrington, David Heller, David Hodge, Stephen Hodge, Jennifer Hyndman, Ben Hyslop, Jane M. Jacobs, Natalie Jamieson, Lucy Jarosz, Susan Jeffords, Andy Jonas, Larry Knopp, Gerry Kramer, Lisa Law, Vicky Lawson, Helga Leitner, Deborah Martin, Kathy Mee, Roger Miller, Don Mitchell, Nina Morris, Rica Nagar, Karen Nairn, Lise Nelson, Eric Olund, Steve Pile, Gerry Pratt, Suzy Reimer, Matthew Rolfe, David Sibley, Janic Slupski, Neil Smith, Matthew Sparke, Karen Till, Anna Tolich, Richard Wartho, Jamie Winders and Craig ZumBrunnen. They have each helped me clarify my often muddied thinking and this book is better for their invaluable assistance. Tom Cooke is to be specially thanked for his help with processing the PUMS data that are examined in Chapter 4, and Darren Smith must be thanked for processing the SAR data as well. At Routledge, I owe debts of gratitude to Sarah Lloyd, Sarah Carty, Ann Michael and Andrew Mould.

This research has been supported through a number of grants and assistantships. I would like to thank the University of Canterbury, Department of Geography, for three very generous research grants. The Department of Geography at the University of Washington must also be thanked for the assistantship it granted me. These monies enabled me to employ a number of terrific research assistants: Colleen Donovan, Samantha Goff-Jones, Sandra Gover, Pete Mayell, Simon Param and Stephanie Prosser. Colleen Donovan must be thanked specially for all of her intrepid and tireless work. Technical support from Michelle Rogan and John Thyne is also gratefully acknowledged.

The Human Rights Commission (HRC) holds the copyright for Figure 1.1, Keith Haring's *National Coming Out Day*, and I am grateful to them for letting me reproduce it here. Specifically thanks to Candice Gingrich and Nancy Buermeyer for that. Portions of Chapter 1 have appeared in a previously published review essay 'Closet Space' in *Society and Space*, volume 14, 1996, pp. 762–70. I thank Pion Limited for permission to reprint them here. An earlier version of Chapter 5 was published in J. Duncan and D. Gregory (eds) *Writes of Passage: Reading Travel Writing* (1999, pp. 185–99) by Routledge and I thank them for permission to expand it here.

Finally, I would like to thank friends who have helped me so much through this project: Dawn Bonnevie, Paul Boyle, Tony Cheong, Marie McRae, Michael Zimmer and Wendy Lawson. Most of all I must thank Jonathan Cronin, whose enduring love still amazes me after all these years and those miles.

Michael Brown<br>
Seattle, Washington<br>
December 1999

# 1

# EPISTEMOLOGIES AND GEOGRAPHIES OF THE CLOSET

## Introduction

'The closet', Sedgwick writes (1990: 71), 'is the defining structure of gay oppression this century.' The boldness of this provocative claim must be understood in relation to the raw power its subject names. The closet is a term used to describe the denial, concealment, erasure, or ignorance of lesbians and gay men. It describes their absence – and alludes to their ironic presence nonetheless – in a society that, in countless interlocking ways, subtly and blatantly dictates that heterosexuality is the only way to be. It is a sign that certainly embodies Foucault's (1980) notion of power/knowledge: that what and how one knows something is always already an exercise of power. Such power has typically been recognised as either homophobia or heteronormativity (Halperin 1995). The closet is an important term in queer theory and parlance because it conveys so simply and efficiently the specificity of oppression based on intersecting norms of gender/sexuality. It allows us to speak our anger and pain about lying, hiding, being silenced, and going unseen. The closet's ontological demands are exacting and exhaustive: that we cannot be in the world unless we are something we are not. And so it not only points to a lack of being in the world, it also signifies the inevitable oppression we face if we 'come out of the closet' either by choice or force.

This little noun does all its important work linguistically, of course. As a part of speech, 'the closet' is a specific kind of figurative slang: it is a metaphor, a term that conveys one meaning through the form of another. But as a geographer I am struck by the fact that we represent gay and lesbian oppression with such a patently geographic signifier. Whatever else it is, the closet is a *spatial* metaphor: a way of talking about power that makes sense because of a geographic epistemology that is largely taken for granted. It is a sign that – often surreptitiously – alludes to certain kinds of location, space, distance, accessibility and interaction (see Knox and Marston 1998). Indeed, it is not hard to come up with spatial aspects of the closet that connect the metaphorical and the material. Consider Figure 1.1, Keith Haring's 1988 *National Coming Out Day* (Marcus 1993). A closet is literally a place where things are hidden. It is typically a small, confining place off a more central, open room. This is depicted in the image by a black interior of the closet placed in contrast to the bold fluorescent green, orange and purple of the exterior room, and the yellow body. By being placed figuratively 'into a closet', gay men and lesbians are marginalised; by coming out, they are liberated. And so we see a jubilant figure in motion, one of liberation and freedom.

Simple enough, right? What would happen, though, if we *spatialised* the closet metaphor? By that I mean using 'space' as a dimension of all social relations by

Figure 1.1 *National Coming Out Day*, by Keith Haring 1988. Copyright HRC, reproduced with permission.

which that power/knowledge gets materialised in the world. What would happen if we made explicit the implicit geographic dimensions of the closet metaphor? We would have to acknowledge the possibility that the closet is not always *just* a rhetorical flourish; that it is a manifestation of heteronormative and homophobic powers in time–space, and moreover that this materiality mediates a power/ knowledge of oppression.

The purpose of this book is to make that very move. It explores the closet's power/knowledge/space at the intersection of its materiality and its metaphor, what Soja (1996) might call a thirdspace.[1] In the chapters that follow, I demonstrate that the closet is neither just a hollow metaphoric signifier nor merely a transit or sign for something else. The argument does not assume a pre-discursive reality either. Rather it calls for a more explicit appreciation of the ways that power/knowledge signified by 'the closet' work because they themselves are always some*where* (and at some scale), and that whereness enables and constrains social relations. It is to move away from using the closet as an ageographic signifier – where its spatiality only *connotes* power; rather than seeing the spatiality as already part and parcel of power/knowledge. Space does not just represent power; it materialises it. Its metaphoric power reflects and reinforces certain aspects of space, which to date remain inchoately and disparately considered. I want to ruminate on the simple, yet often forgotten, fact that certain spaces and spatial relations do conceal, erase and deny – though in ways more multiple and complex than the closing of a closet door implies.

So this book attempts to reorient a tendency in queer theory to conceptualise the closet as an aspatial force. Certainly the spatialisation of the closet remains largely unexplored. Consider synecdochically a sample of queer praxis. Sedgwick's (1990) path-breaking text *Epistemology of the Closet*, for example, argues that the closet is the fundamental axis of oppression for gay men, setting the metaphor as our central object of investigation. Historians like Chauncey (1994) *inter alia* have meticulously broken down the closet that has kept us from knowing the lives of queer subjects from the past. Activists like Signorile (1993; 1995) have insisted on outing famous people from closets within politics, the media and popular culture. Therapeutically, self-help guides are replete with admonishments to come out of the closet (Osborn 1996). Queer praxis (embodied in theory, activism and therapy) has repeatedly stressed the importance of the closet to gay and lesbian oppression, certainly. Yet for all this breadth of attention, these epistemologies remain (as I will argue below) simply metaphorical. Sedgwick and Chauncey, for example, anchor their accounts of the closet in the etymology of the term, underscoring a theoretical distinction between the power it denotes and the spatiality it connotes. Signorile and Osborn are more explicit, stating that the closet is merely a metaphor for oppression. The closet is just a rhetorical device, a literary expression, a signifier for something else. In large part, I think, this perspective is unsurprising given the geography of queer studies themselves, anchored in the arts and humanities and the expediencies of political activism. So here I join Natter and Jones' (1993: 198) call for an increased interaction between geography and literary theory:

> In its critique of representation literary theory has come to understand power as central to the understanding of communication. Yet space in literary theory has been notable mostly for its absence, rather commensurate with its designation as mere metaphor. Space, however, is more than metaphorical, or put differently, 'metaphor'; is itself spatial never empty or nonideological, (social) space undergirds all practices of production and reception, beginning with the materiality of the book itself.

An exclusively metaphorical epistemology also ironically denies recent debates that have emerged at the nexus of the discipline of geography and social theory. The flourish of spatial metaphors within cultural studies generally has left geographers quite ambivalent. On the one hand there is excitement over the increased recognition of spatiality and its significance for social processes (Soja 1996; Massey 1993; Gregory 1994). These geography lessons seem to be diffusing rapidly across the arts and social sciences.[2] Yet even within queer studies space is often relegated to the metaphorical (Browning 1996; Johnson 1996). On the other hand, however, a slew of geographers has expressed reservations about the premises about space embedded in the use of these metaphors (Barnes and Duncan 1992; Pratt 1992, 1998; Smith and Katz 1993; Massey 1993; Price-Chalita 1994; Kirby 1996; Merrifield 1997). The increased appreciation of the inherent discursive nature of social life has found geographers searching for ways to recuperate an interest in 'the real' or 'the empirical' that appreciates the inevitably *recursive* relationship between the metaphorical and the material (Rose 1996; Pile 1996; Pratt 1998). Exploring the materiality of the closet in the chapters that follow, then, balances out and extends our current epistemology of it more generally.

I will begin with a review of how lesbian and gay studies generally, and queer theory specifically, have conceptualised and operationalised the closet. This narrative is structured in 'Epistemologies of the closet metaphor' by a discussion of the central theories of metaphor. I show how the closet operates across various theories of metaphor from the most frequent simple comparison model to more complex poststructural turns. The purpose of this review is twofold. It demonstrates that, despite a variety of theorisations, the closet's spatiality has either been neglected or treated as merely incidental. The diversity also suggests the need for a wide variety of empirical and social-theoretical domains through which to explore its materiality, which is answered in the empirical chapters that follow. Geographic and literary discussions of spatial metaphors are reviewed in 'Epistemologies of spatial metaphors' in order to address geographers' potent concerns over static metaphors of space, and to suggest ways in which an empirical investigation of the closet might allay those critiques. In 'An introduction to the case studies', I introduce those case studies, explaining how and why they were chosen to consider the spatiality of the metaphor.

## Epistemologies of the closet metaphor

### *Etymology*

Taking the word literally, 'closet' appeared in Middle English sometime between 1150 and 1500, and originally referred to a small private room used for prayer or study. By the early seventeenth century it referred specifically to a small room or cupboard, while later that century it was resignified to connote 'private' or 'secluded'. It is probably derived through Old French as a diminutive of *clos* (enclosure), from the Latin *claudere*, which means 'to shut' and *clausum*, which means 'closed space' (Barnhart 1995: 122–3). Standard dictionary definitions mark it as a noun meaning a small or private room, a cupboard or recess. As a transitive verb, it means to isolate, hide or confine something. As an adjective, it suggests secrecy, covertness. These latter definitions, of course, are usually treated as slang. A 'closet queen', the *OED* tells us, is 'a man who is homosexual but does not admit the fact' (p. 287).

The timing of the metaphor can be linked to the rising popularity of its material signifier in domestic space. Busch (1999) claims that as part of the average home the closet grew commonplace in nineteenth-century domestic architecture, then became ubiquitous with the mass-consumption hegemony of Fordism, where an increase in consumer goods demand increased storage space. The subsequent origins of a sexualised metaphor based on domestic architecture remain unclear, however. Chauncey's historiography of gay New York up to the 1940s notes that the term 'closet' did not seem to be used to describe hidden sexuality. He argues that a closet emerged later in that city, and submits that the term itself was coined only in the mid- to late 1960s. Confirming this general history, Barnhart (1995) suggests closet came to mean hidden, covert, or secret sometime around 1968. Interestingly, however, Beale (1989) documents the use of the term in Canada during the 1950s. And speculation abounds as to whether the metaphor emerged from the British 'water closet' to connote cottaging or the expression 'skeletons in the closet'.

Regardless of its precise origins, it is clear that between 1968 and 1972 the term came to signify the concealment and erasure of gays and lesbians specifically in the US. By 1970 the slogan 'Come out!' was a rallying cry in the nascent gay liberation struggle in New York City.[3] Just two years later it was familiar enough to be used in a political book title (Jay and Young 1972). What is interesting is the recentness of this signification towards sexuality (e.g. Chapman 1986; Renton 1995). Since then, the closet has become a 'dead' metaphor (Beale 1989; Barnhart 1995). Dead metaphors are words or phrases that become so widespread and popular that the newness, novelty, or oddity is largely forgotten (Lakoff 1987; Radman 1997).[4] The rapid diffusion of the closet metaphor into gay culture and then beyond is striking insofar as its popularity suggests that an interrogation of the term does indeed have value for social theory more generally.[5]

So if the closet is a metaphor, just what is a metaphor? That term is usually understood as an implied comparison between two things. It derives from the Greek *metapherein*, and literally means to carry over or transfer. Metaphors substitute meaning from one thing to another (Jakobson 1990). Simplicity seems to end here, however, because over the centuries there has been an extended scholarly enterprise dedicated to understanding just how it is that metaphor actually works. This scholarship is impressively interdisciplinary, ranging from linguistics and literary theory to philosophy, aesthetics, and even more recently computer science. It also organises taxonomies of metaphor in multiple and often contradictory ways (Hawkes 1972; Ricoeur 1978; Sebeock 1986; Malmkjaer 1991; Bright 1992; Asher 1994; Crang 1998). In what follows, I introduce three rather broad theories of metaphor to canvas how queer studies have conceptualised the closet. Each is not without its criticisms and tensions with the others, but the outline provides a framework that shows that, despite the wide range of epistemologies, the closet metaphor's spatiality has largely gone unconsidered.

### Comparison theory

Aristotle (1965; 1991) was the first to propose the comparison view of metaphor.[6] He argued that metaphor was just an implied simile, which is a statement of tacit comparison between two nouns, where 'like' would make it explicit. Direct comparisons, with the adjective 'like', are more precisely called similes.[7] Metaphors are uncomplicated transfers of meanings based on some sort of similarity (Murray 1972). Comparison theory presumes there can be a ready substitution between the two items being juxtaposed, and furthermore that metaphors can only work at the sentence level between nouns. For Aristotle, metaphor had no claim to a positive meaning; it worked by subversion (and this point is extended in interaction theory; see below). It was essentially a charming and decorative form of speech circumscribed by poetics (Gordon 1990). James (1960), for example, insists on a distinction between poetic and scientific language, where only the poet's imaginative vision is conveyed by metaphor. Thus, according to classical comparison theory it not an essential, or even ubiquitous, component of language, merely an artistic flourish or embellishment.[8] As Hawkes (1972) notes, this theory of metaphor presumes a straightforward, uncomplicated separation of the world and representations of it and that the manner in which something is said does not alter what is said.

So how exactly are these axes of oppression like a closet? Duncan (1996: 138) provides a useful starting point when she notes that 'The spatial metaphor of the closet is a particularly telling one in this context [of domestic space] where gays may not be "out" even to their own families within their own home.' A closet is obviously a space: typically small and dark and bounded.[9] While small, closets are not as small as cupboards or cabinets, suggesting they could accommodate a human body, though it would certainly be considered odd for someone normally to occupy the space of a closet (Wielgosz 1996): It is a space where things – not people – belong. Yet it is a belonging of a certain kind, for spaces, like closets, contain secrets. Its *location* and *distance* suggests proximity to some wider (more important, more immediate, more central) room, but it's a certain kind of proximity: one that limits *accessibility* and *interaction*.[10] The ubiquity of gays and lesbians 'everywhere' means that on the one hand they are indeed close at hand, but enclosure of the closet means that they are separate, hived off, invisible and unheard. Moreover, as Bachelard (1994: 78) reminds us, it is a space within the private sphere that has a limiting property: 'a space that is not open to just anybody'. The closet is not far away from the room, and it is certainly accessible, but one must look for it. One must open its door to see its contents, or to move into or out of it. So by definition a closet has a certain kind of spatial interaction with its room. It is separate and distinct too. It segregates, it hides and it confines. Closets are spatial strategies that help one arrange and manage an increasingly complicated life (Busch 1999). Simple definitions of 'the closet' in queer culture are underpinned by comparison theory in all these ways. Consider the following explicit definitions from popular gay texts:

> *Closet*: The place where gay men or lesbians hide, figuratively speaking, if they do not want their homosexuality to be known.
>
> (*Alyson Almanac* 1993: 87)

> *Closet*: The confining state of being secretive about one's homosexuality.
>
> (*The Gay Almanac* 1996: 84)

> *Closet*: Term used to describe a state of conscious overt, tacit, or implicit denial of being primarily attracted to one's own sex, as in the expressions 'in the closet', 'out of the closet', 'closet case', 'closeted',etc.
>
> (Hogan and Hudson 1998: 140)

Likewise, the architect Betsky (1997: 16–17) epistemologically knows the closet through comparison theory. He emphasises its location as already in the private sphere, as well as an inappropriate place for a human being to be physically. He also plays on the psychoanalytic aspects of the closet: its ability to compartmentalise one's various subjectivities and desires:

> What is the closet? It is the ultimate interior, the place where interiority starts. It is a dark space at the heart of the home. It is not a place where you live, but where you store the clothes in which you appear. It contains the building blocks for your social constructions, such as your clothes. The closet also contains the disused pieces of your past. It is a place to hide, to create worlds for yourself out of the past and for the future in a secure environment. If the hearth is the heart of the home, where the family gathers to affirm itself as a unit in the glow of the fire, the closet contains both the secret recesses of the soul and the masks that you wear.

Comparison theory also makes more complex claims about the closet, however. It assumes that there is always a liminal space in the doorway. It suggests a possibility for movement and spatial interaction with some broader environment. In other words there is mobility between closet and room. This interaction also has been stressed theoretically. Savran (1996), for example, makes the point that the normalcy of heterosexuality depends to an extent on the very presence of homosexuality as other. In other words the centrality of the room is premised to an extent on the architectural marginality of the closet.[11]

Comparison theory highlights how spatiality is readily part of our epistemology of the closet. While this theory might seem too obvious to be considered at length, it is its simplicity that makes it so powerful and popular. It directly asks us to understand oppression though imagining a common, everyday kind of space. Most importantly, it tropes on meanings of concealment, elsewhereness-yet-proximity, darkness and isolation, with the potential for movement or escape. It anticipates the current parlance of understanding identity through spatial metaphors, a move that presumes that who we are is best conveyed by saying *where* we are (Hetherington 1998). The point I would stress here, however, is the poetic attitude of comparison theory towards space. Ultimately all space can be is a rhetorical bridge, a means to understanding something rather more abstract and fundamental. It is not part and parcel of the oppression we seek to understand, merely a poetic allusion to its workings, which must be something other than spatial. This point becomes more acute in interaction theory of metaphor.

### Interaction and speech-act theories

As hinted at above, metaphors are not just simple direct comparisons between things, because they are often meant to spark a rather ineffable insight or understanding that is difficult to produce with literal similes. There is insight and truth in poetics! Thus one long-identified problem with the comparison theory is that it obscures important differences between similes and metaphors. Metaphors, in one sense, are always lies. According to interaction theory, they work not because they are literally true, but because they denote only limited or partial truth. Extending this point, there must be some sort of interaction between the two terms in order for the reader to 'get' the metaphor. This leads to the *interaction theory* of metaphor, which stresses the conceptual role of metaphor (Waggoner 1990). Here metaphor operates through a semantic interaction between literal and metaphoric levels, and between certain aspects of meanings between two signs.

Early on, Richards (1965 [1936]) identified several elements of metaphors that informed this perspective. The *tenor* of a metaphor (sometimes also called its focus or principal subject) is the general drift or the underlying idea being communicated through a metaphor. The *vehicle* of a metaphor is its source or base, its 'surface meaning' (sometimes referred to as the subsidiary subject) (Barfield 1960). It is the basic analogy that carries the tenor. For Black (1962; 1979), the relation between the frame and the focus of a metaphor was the pivot on which it performed its task. He emphasised the simultaneous similarity and tension between the frame of a metaphor and its focus. In other words, interaction theory proposed that metaphors do not simply function on the page of a text; they help us to understand meaning and language. The similarity is already explained by the comparison theory, but the tension suggests something quite different, yet simultaneous. Metaphors produce a certain shock, surprise, or pause for the reader because of the clash of literal meaning (Beardsley 1972). In other words, it stresses the twist, tension or opposition as well as the easy comparison being made in a metaphor through what Ryle (1955) called a category mistake. It is not simply that *a* is like *b*; *a* is also *not* like *b*. The reader must sort out the similarities and differences being carried by any metaphor, according to interactionists. This implies a complex and invested model of reading, since readers must work both to recognise *and* decode the truth and meaning of a metaphor.

Interaction theory furthermore situates the production of meaning in a thoroughly social context, since reader and writer must share common systems of meaning.[12] It is manifest in speech acts,[13] which are simply acts of communication (Searle 1969). According to *speech-act theory*, metaphor works because of a co-operative recognition shared by author and audience of the difference between 'utterance occasion meaning' versus 'timeless utterance meaning' (Grice 1989).

The former relies on the context of the speech act for its intended meaning. Speech-act theory, then, also places a high stress on the social and cultural conventions and mores that provide a set of common ground rules through which people can jointly – and successfully – make sense of their worlds through metaphor (Searle 1979).[14]

We can read the interaction theory of metaphor also through queer studies of the closet. The *tenor* of the closet is the concealment, erasure and denial of homosexuality in a broader punitive context of heteronormativity. The *vehicle* is a certain kind of space. It is a dark, small, confined place. If you were in a closet, you would probably not be comfortable, and people outside, in the room, would probably not know you were in there, despite the fact that you would be very close and accessible to them. It is also a space where people do not belong. Through the decoding process, readers are left to figure out just why you are positioned so ironically. The spatial locations and interaction between closet and room stands for the social interaction between public and private spheres.

There are several (literal) ways heteronormativity or homophobia are not like a closet: they move, change shape and form, or they can be one aspect of a person, rather than framing the entire subject. These metaphoric mistakes are often taken up in order to probe the complexity of concealment, erasure, or oppression (see below). In the self-help literature, for instance 'coming out of the closet' is a recurrent counsel (e.g. Osborn 1996: 17–18, 21).

> The closet is unique. This institution dominating millions of people's real daily lives is, in fact, a metaphor – a metaphor for denial and hiding. Although traditionally associated to some extent with the notion of privacy, the closet is a fascinating and extraordinarily potent force that is, quite simply, a structured lie. Coming out of the closet then, shatters the lie – breaks silence – and begins the truth-telling that leads to the possibility of a sane and happy life.

> Every gay man, lesbian, and bisexual knows all too well the contours of the closet. And we know that the closet can be a fluid and changeable thing.

> The closet is a shape-shifting phenomenon; the 'out' spheres broaden over time, the closet shrinks.

In each quote above, there is a clear emphasis on the 'category mistake' of using the noun 'closet' to talk about sexuality. These sentiments are echoed across queer studies. Sullivan (in Bawer 1996: 143) notes that 'there is no such thing as a simple closet', which is ironic because the closet is architecturally a rather simple kind of space.

We can also see the interaction theory at work in especially spatial ways in the arguments of both Signorile and Chauncey. In his provocative manifesto for out-ing and coming out of the closet, Signorile (1993: xv) argues that gay America can be mapped by what he calls 'the Trinity of the Closet'. He actually charts three geographic closets associated with distinct power structures in the country. In Los Angeles, he argues, there is a closet around the entertainment industry that suppresses healthy images of gay life out of popular culture such as TV and film. In New York, there is a closet produced by the media industry that ironically delves deep into gay culture but usually represents only the most pathological aspects. In Washington, DC, there is a closet produced by the political system that makes it next to impossible to legislate for gay-friendly issues and bills. For Signorile these three closets are indeed spatialised, but in an interactive way. The spaces are simple *synecdochic* containers for exercises of concealment and erasure. Not everyone in these three cities is closeted by those specific power relations, obviously; nor is every oppression in each city symptomatic of the closet for which it stands. Signorile knows that and we know that. But as an interactive metaphor, the closet's architectural allusions are grounded in places that are conveniently made to hold certain types of oppressions over others.

Chauncey's geographic imagination is similar yet slightly different. He pains-takingly reconstructs a gay urban culture in New York City between 1890 and 1940. His argument theoretically insists that there was not so much a 'closet' as a world. It pivots on a metonymic substitution based on both location and spatial scale, where gay life is not like a *closet*, but is like a *world* (Brown 1996; Jakobson 1990). More importantly, however, the logic of his entire argument rests on both a comparison and interaction theory of the closet. Gay life could either be like a small, confined, private space of storage and concealment, or it could be like the world: a space of enormous size and scale. It was a place of publicity and open-ness where interaction was typical if not inevitable. Either way, gay life (and the oppression inevitably circumscribing it) has certain similarities to space and not others. Readers' ability to understand Chauncey's argument is premised on interaction theory: we know that gay life wasn't literally in a closet or in the world, but he trusts we can understand what he means when he uses those spaces to describe homophobia and heteronormativity.[15]

Interaction theory, then, builds on and extends the simplicity of comparison theory. While in several ways homophobia and heteronormativity are indeed like a physical closet, literally they are not, and in several important political ways they are not. For some, this torsion has meant a dismissal of the metaphor overall, for ones that compare tenor and vehicle more harmoniously. Nevertheless, for all perspectives the nature of space is still allusional rather than material. It is exemplary of heteronormative or homophobic power/knowledge constellations, rather than constitutive of it. In other words, all these takes on the closet are premised by the assumption that space is but a heuristic to understand power/ knowledge, rather than the assumption that all social relations are always already spatial. Oppression is either like or not like space rather than being worked through spatial relations themselves.

### Poststructuralism and metaphor

Poststructural approaches to metaphor show us the oversimplifications of comparison and interaction theories. They have effects on the closet and the entire project of representation that are deep and wide. A thorny and recurrent problem for Enlightenment thinkers was the ubiquity and recurrence of metaphor in thought and speech. Nietzsche (1972) was one of the first thinkers to recognise this fundamental problem of logocentrism and representation that would inevitably lead to present-day postmodernism. Truths, he argues, are simply metaphors that we have forgotten are metaphors. By insisting on the inevitability and inescapability of metaphor, Nietzsche begins to recognise the crisis of representation. No reality can be known or understood without being filtered through linguistic and discursive structures whose independent effects often go unrecognised, and when recognised, can only be understood circularly, never independently of those very templates. Whereas structuralists confidently assumed a stable relationship between signifier and signified, poststructuralism noted the transitory nature of that fixity (Eagleton 1983). Nietzsche's legacy is that questions of how metaphors operate linguistically or cognitively become questions of epistemology and philosophy generally. Derrida (1982), for example, insightfully argues in 'White mythology' that ultimately there can be no such thing as 'metaphorology' because one cannot erase metaphors from any language through which we could study that phenomenon. The ubiquity of metaphor denies any possibility of *retrait* (Derrida 1978). There can be no proto- or pre-discursive language that escapes metaphoric moves. As such, the whole project of comparison and interaction theories is naive because their own explanations are replete with metaphoric understandings of what they are and how they work. Specifically for philosophy (as a process of communication and representation) it renders impossible a meta-language that could epistemologically guarantee the certainty of representation, thought, or truth itself standing outside the vagaries of human subjectivity and cultures (Leddy 1995; Johnson 1995; cf. Ricoeur 1978; Indurkhya 1994; White 1996). Consequently, an important tactic of deconstruction has been to expose the metaphorical power of words and terms nested in philosophical arguments (Derrida 1972). More pragmatically, this point has been pressed by Lakoff and Johnson (1980), who make links between lived experience and broader conceptual systems of thought and understanding. They argue that metaphor has a mimetic relationship to social belief systems and unconscious thought, and therefore metaphors offer us insights not simply into linguistics or rhetoric, but a culture itself. Metaphors are symptomatic of culture (see also Vervaeke and Kennedy 1996).[16] Yet even this relationship can be unstable, according to poststructuralism.

Poststructural theories of metaphor also underwrite the closet trope. Queer theory especially is at pains to stress the paradoxical meanings of the closet, denoting the instability around the meaning(s) of the term. This point is built upon to register the complexity of how homophobia and/or heteronormativity operate. Fuss (1991: 4), for example, insists on an ironic geography of identity politics by claiming that 'the first coming out was also simultaneously a closeting'. To come out, in other words, one must still build a closet. Although poststructural takes on metaphor abound, the most sustained theoretical discussion on the closet is Sedgwick's (1990) *Epistemology of the Closet*. As one of the founding texts of queer theory, the book marks an important starting point in the interrogation of how we know and represent gay and lesbian lives. Sedgwick explores nineteenth-century American and English literature to understand the workings and machinations of the closet. She argues that the closet is a special kind of epistemology for a specific kind of oppression. In her fascinating analysis of the Book of Ruth, she notes how sexuality is a kind of difference distinct from that of ethnic identity. Its uniqueness, she argues, lies in its capacity for secrecy. The closet, therefore, is an 'open secret', a 'knowing by not knowing'. Her flexible, paradoxical language hinges on the liminality of the closet's doorframe.

> The gay closet is not only a feature only of the lives of gay people. But for many gay people it is still the fundamental feature of social life; and there can be few gay people however courageous and forthright by habit, however fortunate in the support of their immediate communities, in whose lives the closet is not still a shaping presence.
>
> (Sedgwick 1990: 68)

> Furthermore, the deadly elasticity of heterosexist presumption means that, like Wendy in Peter Pan, people find new walls springing up around them even as they drowse: every encounter with a new classful of students, to say nothing of a new boss, social worker, loan officer, landlord, doctor, erects new closets whose fraught and characteristic law of optics and physics extract from at least gay people new surveys, new calculations, new draughts, and requisitions of secrecy or disclosure.
>
> (Ibid.: 68)

These are arguments that insist that the closet's power is more complex than its metaphor immediately recognises. The ironic turns in this logic are premised on the insights of poststructuralist literary theory: that there is never an outside to representation, and that meaning is generated on absence as well as just presence in a text.

Poststructuralism also frames debates over whether the closet is a good or a bad thing in gay life. This signification is witnessed in debates over outing in the mid-1980s (where for some the closet wasn't just oppression, it was also a place of safety or individual privacy to be respected) (e.g. Warner 1994; McCarthy 1994; Gross 1993) and more recently in the so-called post-gay movement. In parody, arguing from a post-gay perspective, LaBruce and Belverio (1998) have argued 'the case for the closet' (cf. VanDecimeter 1999). For them, the closet has been a place of excitement and excellence. Doing away with the closet relegates gay culture to an abject mediocrity. What is interesting is that post-gays theorise the closet precisely not as a fixed, confining space, but rather as a secret and effusive, ethereal influence. Its meaning has become completely antonymical:

> *Judy:*   I think that homosexuality should be an invisible influence in culture. It shouldn't be something that's ghettoized or flaunted it should be an invisible influence on style, fashion, the media. . .
>
> *Glennda (helpfully):* Cinema.
>
> *Judy:*   . . . and that is its strength. If it's confined to a specific identity or to a specific geographical location or any kind of ghettoization, it becomes watered down and leads to mediocrity.
>
> (LaBruce and Belverio 1998: 152–3)

For Glennda and Judy, the space of the closet is everywhere to the extent that it is nowhere. The metaphor has been reinterpreted beyond its familiar spatial connotations as its very meaning is resignifed.

Despite the productive insights generated in this strand of thinking, poststructural takes on the closet metaphor also elide the spatiality of homophobia and heteronormativity. Largely this is due to the stress placed on deconstructing the hegemonic simple, taken-for-granted meaningfulness behind signs and representations (like the closet). This hegemony has been assiduously deconstructed by queer theorists as part of a larger critical project that shows us just how intricate the workings of power/knowledge can be. Such a tactic focuses on textuality, however, rather than spatiality. Yet even texts themselves have spatial dimensions to them.

Poststructuralism's explanation and use of the closet metaphor has a number of implications for the chapters that follow. It notes the oversimplification in comparison and interaction theories. Speaking of 'the closet' is not simply a rhetorical or poetic flourish, nor is meaning as straightforwardly produced as it would seem. A poststructural theory of the closet metaphor rejects any simple prioritisation or stable relationship of the empirical frame over the theoretical tenor, or between the signifier–signified in the closet-sign. It implores us to be aware that metaphors can carry along with them a whole system or networks of beliefs that do powerful epistemological work, but remain tacit and unacknowledged. Additionally, we can never forget that discourse, and the relentless metaphoric strands that web it together, is always inescapable. So assuming a clear-cut distinction between the metaphorical and the material is specious. We can never get outside the closet metaphor; we can only resignify it, or understand it with yet another metaphor. Most importantly, we can never use 'the real world' to guarantee an authentic, true, or essential meaning to the closet.

For poststructuralists, it is not simply that certain allusions of the sign 'closet' are true and others are false, it is that the very stability upon which comparison or interaction theorists could make such a claim can inevitably be subverted. The fact that the closet can refer to a space does not anchor it into any sort of spatiality, for it can be resignified, and clearly the deadness of the closet metaphor is evidence of this very process. My point, however, is that even from poststructural perspectives the closet's epistemology is only rarely spatialised. And when it is, it is done in a narrow sense: that of the text itself rather than empirical social settings. Sedgwick's work, for instance, is primarily a textual geography of the closet. Even when she considers the sheer materiality of home, the closet is not spatialised (Moon *et al.* 1994)! This is not to deny the inherent sociality of the text, or the discursive inevitability of 'the real world'. It is to suggest, however, that other modes of investigation, ones that are more conventionally 'geographic' in the sense of having a terrestrial location, might also provide insights into how we know the closet, and the import of that knowledge for how we know society. Acknowledging the slip and sway of meaning in a spatial metaphor does not necessarily imply rejecting any consideration of the spatiality of the processes it signifies. Indeed, if anything, it makes such an intellectual move all the more possible and important, lest we only come to know the epistemology of the closet through literary texts principally.

By reviewing these three rather broad theories of metaphor generally, we can see how the closet as a sign works in both simple and complex ways to name exercises of power/knowledge that oppress. We can recognise the intellectual work so tacit in our own understandings of 'the closet'. What also becomes apparent is that all these perspectives carry rather aspatial conceptualisations and operationalisations of the closet's power. They skirt the possibility that the closet is not just a metaphoric move to describe a social process. Even Signorile's geographic allusions merely use cities synechdochically, for the sake of a streamlined argument. Understandings of the closet in queer theory to date do not treat space as if it were constitutive of social relations; rather they take it to be merely representational of them. Closet space is a poetic licence or an easy substitution based on straightforward similarity according to comparison theory. Interaction theory asks us to sort out the ways oppression is and is not like a physical closet, but the premise remains that it is not spatial in and of it self. Poststructuralism troubles us into acknowledging the lack of stable or fixed meaning to the sign 'closet', thereby suggesting that any appeal to the 'real world' that space and geography imply is inevitably intellectually sophomoric. Literature on spatial metaphor specifically, however, provides a means by which to take the spatiality of the closet seriously, without ignoring the serious challenges poststructuralism offers. A review of that literature also deepens our understanding of just how the closet metaphor does its signification, and why it is so popular.

## Epistemologies of spatial metaphors

Girding the literature on spatial metaphor are three themes that also inform my call for a consideration of the closet's spatiality. At a preliminary level is the recognition of the closet as part of a vast array of spatial metaphors in English generally. The sheer ubiquity of such metaphors has long been recognised. Their omnipresence is reflected and reinforced by the plethora of disciplines and per-spectives from which scholars have investigated them: from linguistics (Radden 1985), literary criticism (Mathur 1996), feminism (Higonnet 1994), film theory (Thompson 1993), religious studies (Noppen 1974), computer science (Trumbo 1998), psychology (Hanenberg 1982), and geography (Smith and Katz 1993). One reason for this span is the simple linguistic point that our language is actually shot through with spatial metaphors at a syntactic level. Dirven (1983), for example, shows us that nearly all prepositions (on, at, up, down, etc.) are spatial metaphors in a sense. 'In a sense' from the previous sentence is just as much a spatial metaphor as 'in the closet', then. These metaphors are also used as a means of reification in order to signal duration, intensities and tendencies (Worf 1956). Spatial metaphors are also effective metaphors because many have a causal 'subsense' that is rhetorically compelling, Radden (1985) argues.[17] Linguistically speaking, it is a rather common structure of signification.

A rather different argument about the ubiquity of spatial metaphors is made about the fact that as corporeal beings we exist in time and space. Several authors here argue that spatial metaphors are popular, useful, or ubiquitous precisely because they reflect our bodily positions in space. Jones (1982) argues that the concept of space is intimately related to consciousness because it tells us we exist as a body separate from, yet related to, things. He calls it a 'cardinal metaphor'. He goes on further to suggest that spatial metaphors imply individuals' very existence and ontology (see also MacKay 1986; Fleischman 1991; Schwartz 1996;

Boers 1996). Behavioural education and psychology perspectives argue the importance of spatial metaphor because of the proprioception grounded in the essentialised experience of inhabiting a body, or because of some inevitable stage of infancy (Tolaas 1991). Clearly the closet works on this corporeal dimension: trapping, hiding, and confining the queer body along the lines of the comparison theory of metaphor.

Explanations for the popularity or ubiquity of spatial metaphors typically turn on their economic ability literally to ground abstract, ethereal or diffuse meanings in texts (Noppen 1974; Meisel 1979; MacAdam 1980; Leeman 1995). Hanenberg (1982) and Roediger (1980), for example, agree that spatial metaphors in dreams (or in their descriptions) provide a dimensionality that augments the psychological work dreams do for us. Ranciere (1994) claims that spatial metaphors allow for a complex interaction between the known and the experience, the text and 'the real world'. Similar arguments are made with respect to the proliferation of spatial metaphors in computer software *vis-à-vis* cyberspace (Kim and Hirtke 1995; Damle 1997; IBM Research Division 1997; Trumbo 1998). A great deal of feminist literary criticism appreciates the possibility of spatial metaphors because they allow authors to convey complex emotional or political dimensions of women's experience with a certain clarity and economy (Mathur 1996; Sample 1991; Philbrick 1979; Mann 1982).

Within social theory spatial metaphors have gained popularity on precisely these grounds. They provide a means to talk about social position and identity in a way that remains contingent, unfixed, but still 'there'. Bondi (1993: 98) puts it well when she argues that 'geographical terms of reference do the work done by essences in other formulations'. Certainly the ability of the closet metaphor to signify both the psychological and social dimensions of alienation resonate with these claims. Moreover, its popularity can be explained by the fact that it has come to describe the simple and complex dimensions of homophobia and heteronormativity. Besides conveying a sense of entrapment simply, it can also connote the complexity of self/other relations: knowing by not knowing, familiarly represented in the liminal space of Fuss' inside/out position.

A second theme in the literature is the debate over whether spatial metaphors are good or bad modes of representation. Following on Bondi's point above, several authors have asked us to consider the cultural norms implicit in certain spatial metaphors, and how they tacitly work to influence texts (for instance, up is good, down is bad, etc.) (Mercken-Spaas 1977; Wallace 1988; Langston 1994). Silber (1995) has criticised the use of spatial metaphors in recent sociological theory as a kind of creeping positivism. At its strongest, she claims it allows sociology to speak scientifically, to reify, objectify, to claim ontological status for the phenomenon under study. Likewise, several authors have complained that spatial metaphors often connote essentialised subjects or simplistic dualisms that are theoretically reductionist and politically problematic (Martens 1982). The often unexamined biases and traces in spatial metaphors have bothered linguists, feminists and geographers alike. Mercken-Spaas (1977) notes an often ignored normative bias around the spatial dualism of inside versus outside. Is 'inside' security or confinement? Higonnet (1994: 197) has argued that spatial metaphors in literary criticism as well as literature have 'a political, classist and racial aura and impact', concerns echoed by Caplan (1998).

This critique has been visited several times in the last decade by geographers who have been bothered by the explosion of spatial metaphors within cultural studies and social theory.[18] They rail not so much against the authors' politics as their inchoate notions about space more generally.[19] In looking at the metaphor of 'space' in Laclau's work, Massey (1993) finds that it is used uncritically and simplistically. I have made a similar critique against radical democratic theory (Brown 1997). Merrifield (1997) worries about a troubling representational dualism in urban theory whereby only temporal metaphors in urban theory convey a sense of process and transformation, while spatial metaphors are deployed to suggest individuation or everydayness, and only static spatiality is conveyed. Perhaps Smith and Katz (1993: 79–80) have been the most worried over the preponderance of spatial metaphors in cultural studies that only conceptualise space in an absolute sense as static, self-evident. In their own words:

> the uncritical appropriation of absolute space as a source domain for metaphors forecloses recognition of the multiple qualities, types, properties, and attributes of social space, its constructed absolutism, and its relationality. This is not to say, therefore that absolute space has no real referent; in modern representations of the body, private property, the state, and colonization, absolute space is very real, if socially constructed. The problem lies rather in the naturalization of absolute space which leads, in turn, to a tendency for such metaphors to become virtually free-floating abstractions, the source of their grounding unacknowledged. The widespread appeal to spatial metaphors, in fact, appears to result from a radical questioning of all else, a decentring and destabilization of previously fixed realities and assumptions; space is largely exempted from such skeptical scrutiny so it can be held constant to provide some semblance of order for an otherwise floating world of ideas.

Yet recently care has been taken not to overgeneralise this critique (Massey 1993; Price-Chalita 1994; Brown 1996; Pratt 1992; 1998). Price-Chalita argues that there is no necessary connection between spatial metaphor and politics. Pratt (1992) agrees, advising the need for ambivalence towards any metaphor, but especially spatial ones. She argues that different sorts of spatial metaphor provide and occlude insight simultaneously. The rhetoric of mobility and exile, for instance, offers a way to disrupt centrality of certain power/knowledges, but also 'runs the risk of reproducing the privilege of the unsituated observer' (1992: 242).[20]

With reference to the closet, I want to build on Pratt's (1998: 26) insight:

> The utility of a metaphor to some extent depends on the specific circumstances in which it is used, and there are no doubt times when a static, simplifying metaphor may be useful precisely because it obscures detail and other viewpoints for the sake of a specific political (and polemical) aim.

Very often it is quite politically useful to name queers' position as in the closet. Its efficacy stems precisely from the accounts of metaphoric theory and spatial metaphor's workings reviewed above. With respect to the dilemma of the closet's simple signification of stasis and fixity, I would argue that it is often very important to recognise this quality of space, even if there are more complex, less straightforward workings of the spatial. Sometimes space does simply confine, conceal, trap and disempower. This is precisely because exercises of power/knowledge that oppress can do their work through spatial constructions or arrangements. Exploring the closet as a spatial manifestation of homophobia and heteronormativity – and not just a metaphor for them – brings this point into high relief.

A third, and most recent, theme in the literature is a growing will to reject the dualism of spatial metaphor/materiality, in light of the poststructural insights discussed above. Rose (1996: 58) notes geographers' inherent drive not merely to prioritise the real over metaphoric space, but in the process to suppose a straightforward division between the two:

> Implicit in these distinctions (and often explicit also) is a hierarchization: real space is understood as a more accurate description of causal processes, and it is therefore more important for geographers to study. For all of these geographers, then, there is a real space to which it is appropriate for metaphors to refer, and a non-real space which it is not.

She goes on to argue (p. 59) that there is an inherent masculinism in the assumption of a clear-cut distinction between material and metaphoric space, where real space is conceptualised as concrete and dynamic, while metaphoric space is fluid and imprisoning. Consequently, she refuses to be pinned down or to choose material and metaphorical ways of understanding or representing space. She tries to write with a deliberate transit between the two, though clearly the text priori-tises the metaphorical over the material, ironically. Likewise, Kirby (1996) does not see one representation as more foundational or explanatory of the other. In her own words: 'space helps us to recognize that "subjects" are determined by the anchoring within particular bodies or countries. At the same time, space in the abstract maintains a fluidity, a revisability that appeals to the reformative impulses of today' (Kirby 1996: 7). Here Kirby argues that one reason for the popularity of spatial metaphors within poststructuralism is because space is treated as simply an effect of discourse (pp. 108–10). She speaks of spatial 'registers' to mediate between metaphorical and material (political, semiotic, somatic and psychic). I am drawn to the possibilities these authors raise for the epistemology of the closet. I cannot, however, completely follow Rose's invective against the priori-tisation of 'real space' explicitly because (as I have already shown) exactly the inverse is the case within queer theory (e.g. Scott 1990; see Jagose 1996). There has been a lack of considered appreciation of the spatiality of all social relations in that literature (Bell and Valentine 1995a; Ingram, Bouthillette and Retter 1997). I am, however, motivated by her more general project to prioritise neither the real nor the metaphorical. In this way, what follows is an attempt to balance and extend. Rose and Kirby suggest the metaphorical epistemologies of the closet can be extended with scholarship that considers the closet as also a materialisation of power/knowledge, as an oppression that works through space, and not just simply through language or texts.

## An introduction to the case studies

Drawing the previous discussions together, I wish to make three imbricated arguments that structure and legitimate the chapters that follow. First, the diver-sity of ways the closet works as a metaphor for power/knowledge suggests that any investigation of the spatiality of the closet must be multiple and variegated. In the chapters that follow this point is translated in two ways. I will examine the closet in a variety of geographic locations across the globe: the US, the UK, New Zealand, as well as travel writing about Asia, South Africa and Egypt. Simul-taneously, I will examine the closet through a nested series of spatial scales: the body, the city, the nation and the world. Given the relational aspect of the closet, the empirical chapters that follow are structured as a series of ever-widening spatial scales in which the closet is constructed. They consider the multidimen-sional exercises of power it has come to name. While this strategy is not without its foreclosures and problematic effects (see Chapter 6), it follows Natter and Jones' (1993: 170) vision that, 'Places result from a spatial "framing" of a particu-lar scale, from the nation-state, to regions, communities, and neighbourhoods, and even to the microsettings within a house.' In this way, each chapter adds to the complexity of understanding to the spatial metaphor, but also allows that specific spatial practice to speak to theoretical debates.

Second, the ubiquity of spatial metaphors generally, and the different ways of appreciating their operations, imply that the epistemology of the closet must be implicated in a wide variety of social theories. Therefore, I will stage dialogues between the closet and a rather wide (though perhaps fenced) range of contemporary critical social theory: Butler's concept of performativity, Lefebvre's arguments about the social production of urban space, Foucault's remarks on governmentality, and Lacanian notions of desire. Third, I recognise and accept the poststructural deconstruction of the dualism between metaphorical and material space. What I would argue, however, is that this need not mean an abandonment of the distinction *tout court*, nor should it mean a one-way transit of insight where the material dissolves into, or is trumped by, the metaphorical. In other words, I do not feel obliged to jettison the empirical domains with which geographers are comfortable to acknowledge the inescapability of metaphor, discourse and textuality. For if materiality is always discursive, so too are texts themselves spatial (just as they are temporal).

In Chapter 2, the relations between the closet and the individual subject are explored through an analysis of published oral histories of British and American gay men. The aim of this chapter is to consider the closet as a performative, which has been a very important concept in both queer theory and geography. The chapter highlights the spatial baggage that notions of performativity often implicitly carry. Here I show how the closet does indeed provide a 'stage' for performativity of sexuality, redirecting some of the spatial attention in Butler's work towards the power of immediate, situational context. I also reconnect her work with its linguistic roots by showing how the closet is a performative in the linguistic sense as well.

Considering the closet as a component of urban space is the task of Chapter 3. There, I offer a textual reading of the inner city landscape in Christchurch, New Zealand. This reading illustrates how the closet for gay men is not simply a function of their sexuality, but is closely linked to the structuration of capitalism in the city. To make this point, however, demands a critique of Lefebvre's widely popular arguments about the production of urban space, which have certainly elided issues of sexuality in spite of their breath and complexity. Here I draw on Harvey's work to connect the widely influential work on Lefebvre with recent work by Knopp that argues the need to see the mutually reinforcing spatialities of sexuality and capitalism in order to understand the effects of each on the city.

How does the closet operate at a national scale? Questions of the relations between nationalism and sexuality are certainly a current theoretical issue. Their empirical formulations, however, have tended towards fictional texts and legal discourses. As a geographer, I have become much more interested in more subtle and geographic exercises of national definition. For this reason, Paul Boyle and I examine the censuses of the United States and United Kingdom in producing the closet in Chapter 4. To what extent can we see gays and lesbians through the census? If we can, what sorts of assumptions have to be made, and how do they compromise the validity of our sightings? Here, we address ideas of governmentality as a way in which sexuality can be seen as an exercise of disciplinary power in statistical exercises of national definition.

In Chapter 5, I examine the significance of the closet to the framing of the world and to theories of desire. Lacan's theory of desire, which argues that desire is a fundamental lack in the decentred subject, is specifically considered. Framings of gay men's desire at a global scale, however, show a closet that is not so much a lack, but a productive if occluded space. Here I examine one gay author's travel writing that has attempted to frame the world – and various closets within it – for gay men. Neil Miller's work, I argue, has important insights for psychoanalytic theories of desire that inform our conceptualizations of queers and the closets that often forcibly 'contain' their desires. It also empirically sustains early queer critiques of Lacan from within French poststructuralism.

## A reflexive conclusion

Let me put the limitations of this work that I recognise up front the better to situate this text. This book is obviously only a partial take on the closet, and that fact is true in several different ways. My theoretical angles are obviously selective and inevitably biased, so these are certainly not the only ways the closet might impact on social theory. In part, the bias stems from theoretical purviews into which several of my colleagues and my discipline have been drawn recently. In part my choices stem from the age-old challenge of articulating empirical and theoretical research. For these reasons, I insist that there is no necessary or essential connection between the theories and empirics I juxtapose in each chapter. In part, my selections reflect (and reinforce) my own intellectual strengths and weaknesses as well as several institutional and practical constraints I faced while doing the research for this book. The closet must be made to work in and with a variety of theoretical contexts to truly queer social theory. Thus I see this text as a fugue: reiterating a search for the closet across a diversity of theoretical and empirical domains. It remains, admittedly, only one attempt.

I also want to be clear that the subjects and social identities who are considered in the chapters that follow are absolutely not the only ones who occupy closet space. Most obviously, I have focused primarily on closets located in gay men's lifeworlds – and clearly only certain gay men at that! I do recognise the problematic closeting effects of this focus for other queers (Jagose 1996; Calhoun 1995; Jeffereys 1994), but also found it too problematic and exhausting a task to speak much beyond gay men's closets. The closet's relationship to lesbians, for example, is only considered in Chapter 4 and part of Chapter 5. This itinerant presence/absence certainly raises the issue that to open one closet door must inevitably close another. To guide my work I have tried to follow Alcoff's (1991) admonition that, while it is often certainly problematic for me (a gay man) to speak for the other (lesbians), it is at times irresponsible not to as well. Thus the closeting effects of the census on lesbians and gay men are probed in Chapter 4 because not using the available data at hand seemed more problematic than ignoring it. Likewise, considering that Miller did speak with lesbians in his travel writing, it would have been disingenuous to avoid those parts of his work. Of course there are other 'others'. Transgendered and transsexual queers are notably missing, as are a variety of certain bi- or pansexual subjectivities. Intellectually, it seemed most prudent to sacrifice breadth for depth in understanding the often confusing shuttle between materiality and metaphor. Besides partiality in the theoretical debates considered here, the case studies offered are no less partial and biased. Obviously there are other, perhaps more important, closets at each of the scales I identify. Public-sex beats at the urban scale, or the 'don't ask–don't tell' policy at the national scale in the American military, come immediately to mind as two noteworthy elisions. The spatialisations discussed here were chosen partly out of my own positionality and locations, my interests in cultural and political geographies, and out of their availability to me as a researcher. Indeed, several closets I researched for this book remain absent from the text partly out of an ethical concern for not outing people against their will. Ultimately I would ask the reader to take the case studies for what they are: a selective yet still telling array of closets from the body to the globe. There are no doubt other closets that tell other tales.

To write about 'the closet', or about 'gays', and to do so in and through empirical material textually fixes certain meanings of those terms, and perhaps this is the most vexing dilemma I encountered in writing this book. It can denote binaries like 'gay/straight' or 'inside/outside' the closet that critical, poststructural theories of language have deconstructed. Fuss (1991: 5), for instance, reminds us, 'The problem, of course, with the inside/out rhetoric, if it remains undeconstructed, is that such polemics disguise the fact that most of us are both inside and outside *at the same time* [emphasis mine]'. She and so many others note that the power we try to resist often works ironically through the very signs we use in our writing. Exploring the spatial metaphor of the closet in gay men's lives through empirical material and in a written text will invariably reproduce fixities around power and identity that conceal and simplify important processes. And so what are we to do? Never write about, never intervene in these struggles? I come to no happy solution here. I have attempted to be careful with my language and the baggage it inevitably carries. Nevertheless, I would point out that, while binaries are in some ways problematic, in other ways they are not. Moreover, such criticisms often themselves deploy implicit dualisms that are dangerous too. So to explain my position, I would point out that I have heard Fuss' other voice (1991: 1) as well:

> The figure inside/out cannot be easily or ever finally dispensed with; it can only be worked on and worked over – itself turned inside out to expose its critical operations and interior machinery. To the extent that the denotation of any term is always dependent on what is exterior to it . . . the inside/outside polarity is an indispensable model for helping us to understand the complicated workings of semiosis.

Here I read her admission that, despite the fixity of dualisms, at times it is quite politically responsible and necessary to work with binaries and dualisms. More-over, it is honest to admit that they are inevitable. For example, I must criticise Fuss' temporal focus in her first quote above that elides the very spatiality this book tries to consider (see Soja 1996).

In summary, theories of metaphor ranging from simple comparison theories to more complex poststructural ones each point to important facets of the workings of the closet metaphor. These theories operate across queer theory, yet do not appreciate the spatiality of the metaphor with a geographic imagination. Older work on spatial metaphor has stressed its proprioception, its significance in orienting the body to the world. More recent work has suggested the novelty of spatial metaphor *contra* the temporal fetish in most twentieth-century social theory. Geographers, *inter alia*, offer several important words of caution in embracing the productivity of spatial metaphor: a synecdochical understanding of space that emphasises only its limiting, confining properties rather than its inter-active, relational and interpellative roles. Still, there is insight in recognising the ways that space does confine, discipline and alienate. The chapters that follow, then, explore the closet both as a material space and simultaneously a metaphor for power/knowledge of gay men. They situate the closet and those it would enclose with geographies that vary by context, scale and location. Let us now open the closet door.

## Notes

1 'Thirdspace' is a notoriously difficult term to pin down, though Soja often uses it to escape the unhelpful and obfuscating dualisms that riddle modern theorising. I draw on it here to signal my attempt to move between metaphor and materiality, and to begin to move beyond that dualism.

2 See edited collections such as Bell and Valentine (1995a) and Ingram *et al.* (1997) for a diverse range of geography lessons for queer studies.

3 The manifesto adorned the Gay and Lesbian Front poster in 1970 and was the title of its newspaper in 1969 (Duberman 1994).

4 A recent issue of *Time Magazine*, for example, used the metaphor to describe the growing awareness of people who self-mutilate (Edwards 1998).

5 Though, as Radman (1997) has argued, there are considerable difficulties in establishing exactly when and where a metaphor dies. He points out just how dynamic and flexible meaning is, in contradiction to the static metaphor of death used to describe the state where metaphors become so popular they lose their metaphorical identities.

6 This is sometimes also referred to as the substitution theory of metaphor (see Asher 1994: 2453). Substitution theory argues that metaphors are 'decorative substitutes for mundane terms where a heightened rhetorical or aesthetic effect is desired, but without cognitive support'.

7 Often similes are treated as simply one form metaphorical language can take, however.

8 For a more nuanced and updated account of work in comparison theory, see Chiappe 1998. He stresses the complexity of information-processing people must do to understand metaphor, which is based on similarity.

9 It is interesting to note that in Seattle, Washington's currently tight rental market, the distinction between a closet and a (bed)room is defined by the presence of a window!

10 In a different context, Abbas (1996: 215) has referred to this relationship as *hyphenation*, 'a space that is both autonomous and dependent at the same time, both separate from and connected to other spaces'.

11 Indeed, this point leads to a more sophisticated appreciation of metaphor, which is taken up later in this section. It also raises the problem of language's fixity and crisis of representation. I address this issue in the conclusion of this chapter.

12 Several problems have been noted with this theory, not least of which is the problematic distinction between literal and figurative language. Davidson, for example, argues that metaphors semantically operate through the literal meanings of their words and nothing more.

13 The term is used to describe the event of communication (indeed, any form of language use, oral or otherwise) in a non-functionalist way (in other words, the function of communication cannot completely explain the existence or all of the effects of speech). People do more with words than simply convey information.

14 Concomitantly, it stresses the importance of the utterance itself as part of the tool kit by which understanding is struck. While speech-act theory acknowledges the interaction between immediate and general meanings in an utterance, it also underscores the importance of performativity, which is a crucial part of the closet metaphor's power. Here speech-act theory would emphasise the power of 'coming out of the closet' by declaring one's homosexuality. Indeed, the ethical debates around 'outing' in the 1980s mark this speech act as a central force of the closet metaphor. This is especially important in theories of performativity; see Chapter 2.

15 Frantzen (1998), who examines Old English literature, stakes a similar argument. He proposes 'shadow' as a better metaphor than 'closet' to describe the positionality of gay men. To claim there is a better metaphor than the closet means the interaction between tenor and vehicle needs improvement.

16 A popular example of this sort of argument would be Sontag (1990).

17 'By', 'of', 'from' and 'through' are specifically noted as signifying causality by Radden.

18 For more general work on metaphor within geography see Jarosz 1992; Smith 1996; Curry 1996; Cresswell 1997.

19 A similar complaint is levelled in film theory by Thompson (1993).

20 Indeed, several scholars outside geography have noted the political possibilities around spatial metaphors in literature (Handelsman 1974). Dreeuwss (1998) sees political possibility in spatial metaphors through their use by transgendered people to discuss the complexities of their identities. Jane Austen's work represents a dilation of space, which reconfigures her characters' growing sense of self (Person 1980). In considering the personal narratives of women academics, Hane (1995) also finds spatial metaphors to be valuable tools by which they understand the relations between home and work (see also Solomon 1989). These scholars collectively argue that spatial metaphor helps to represent and understand the decentred subjectivity of the subject, around the dimension of gender especially.

# 2

# THE BODIES IN THE CLOSET

## Performativity, space and personal narratives

## Introduction

The Seattle Metro bus no. 7 stopped abruptly to pick up two very wet people just at the crest of Capitol Hill on a rainy Tuesday afternoon. The sudden braking caught everyone's attention, and broke the passionate soul kissing of a man and woman sitting just across from me. Since I was sitting towards the rear of the dingy bus, I had a long view of a slender, trendy woman making her way purposefully down the aisle. Behind her, I heard her companion before I could see him. We all could, because he was speaking so loudly. With a mixture of aplomb and hubris our new rider proclaimed, 'That's right, people, I'm swinging my hips as I walk on by. And if you don't like it, you can kiss my beautiful queer ass!' With regal camp he sashayed down the aisle, past my seat, never once breaking his stare forward. On the other side of the aisle, the young heterosexual couple 'tsked', huffed and 'Oh, Gawwwwd'-ed this young gay man just audibly enough to make their revulsion clear to those of us in the back of the bus. 'Who said that?' the gay man demanded loudly.

Everyone on the bus began to grow visibly uncomfortable. After all, this was Seattle. 'I did,' the woman stated loud and clear, but without turning to face him. Then she whispered something inaudible to her boyfriend and they both laughed. 'Well if you don't like it, girlfriend, *what the hell you doin'* up on Capitol Hill in the first place!'

For me, this little slice of urban life foregrounds bodies and closets when sexuality is conceptualised as a performative. It does so, however, precisely because of – rather than in spite of – a complex spatiality of this speech act. Judith Butler (1993a, b; 1997a, b) introduced 'performativity' into social theory to highlight the social constructedness of gender – to move critics and apologists away from the premise of essential, a priori definitions of 'gender' and 'sexuality' (Butler 1993b; Sedgwick 1993). By emphasising the triptych of repetition, iterativity and allegory, Butler insists that one's gender is not simply fixed or given, but that it's fashioned from the discourse and culture around us. It is a structuration that in itself is an exercise of a power(s) with which we are (at least partially) complicit, no matter how intentionally we try to resist. And because everyone 'does' gender all the time, though never exactly the same way twice, it seems natural, pre-given, fixed and essential.

Back on the bus, the heterosexual couple certainly felt comfortable enough to portray their gendered sexuality by kissing in an urban public space, a location they seemed to feel sustained heteronormativity. The gay man likewise signalled – even if it was just a strike at resistance – his gendered sexuality through a camp performance. But it was more than that. It was a performance of a certain kind: that of being *out of the closet*. It was the immediate space of the bus that challenged his presence: his performativity was so transgressive that a heterosexual felt bold enough to shatter the anonymity and atomisation of bus culture to try and shame him audibly. But the event did not end there because the situation of the bus in a gay neighbourhood paradoxically worked to defend his position out of the closet too, didn't it? He would not be closeted in the place through which the bus travelled, through which it interacted. It was the straights who had transgressed on Capitol Hill, a gay neighbourhood in a gay-friendly city. There homonormativity obtained, and breeders should keep their hateful comments in the closet![1]

What is fascinating to me is that work on performativity to date has made it difficult to envisage the closet as anything but a dead metaphor. Queer theory has not self-consciously considered the relationship between the closet (as a metaphor for sexual subjectivity) and performativity (as a textualised understanding of the same thing).[2] If it did, it seems, it would probably emphasise the repetition of this kind of scene elsewhere, or the citational practice within the boy's camp, or the iterative use of shame in the couple's reaction – rather than the geographic contexts (read: locations, space and places) of these performatives. I argue in this chapter that both metaphoric and material dimensions to the closet provide important geography lessons for Butler's notion of performativity, and by extension Sedgwick's own theory of the closet.[3] In order to speak to those theories, I employ a content analysis of 125 gay men's published oral histories from both the United Kingdom and the United States. I consider how they conceptualise and operationalise the closet in their narratives in order to infuse performativity with a geographical imagination: one that tacks between metaphorical and material closet space for the individual body. The gay men's narratives, I argue in this chapter, pull performativity in two directions. First, they illustrate how the closet is often materialised in empirical contexts that are significant to the performance of sexuality. This point extends recent geographic work on the importance of context to performativity. Second, the closet-as-metaphor may be thought of as a performative itself in a linguistic sense. If a performative speech act is a 'doing by saying', then it seems patently obvious that coming out or staying in the closet is usually materialised in the form of a speech act.

Towards that end this chapter progresses in three steps. In 'Performativity and context', Butler's theory is outlined with particular reference to its roots in Austin and Derrida.[4] From these origins, a deep suspicion of the epistemological force of 'context' is highlighted and explained. Butler's Althusserian tendencies emphasise the ethereal interpellation of gender and sexuality,[5] rather than an agent's actions in specific contexts and performances. This point is criticised in the section 'Ignored geographies in performativity', by noting the ignored geographies in 'context' and elucidating the implicit stage metaphor in performativity theory. Here the work of geographers is helpful. Geographers have stressed recently the significance of context to performativity's arguments. Yet, ironically, they have been especially suspicious of this metaphor because it invites aspatial conceptualisations of social life. In 'Speaking of the closet: personal narratives', this impasse – which Butler's work cannot presently recognise – is broken by drawing on gay men's narratives and oral histories to produce insights about the closet metaphor. It can be a context for performativity, as well as a performative itself.

## Performativity and context

As a professor of rhetoric and literature, Butler draws on Austin (1975) and Derrida (1982) to develop her notion of performativity. Austin introduced the term to literary theory in order to describe speech acts that 'did things'. He was getting at the point that sometimes the spoken or written word doesn't simply exist, but often performs some sort of task by virtue of its presence, audibly or textually. He called these signs *performatives* and distinguished them from *constatives*, which are utterances that are simply assertions or statements of fact. For example, 'I christen this child . . .' or 'I forgive you' or 'I now pronounce you man and wife' are all examples of performative speech acts. Performatives cannot be judged on the basis of truth or falsity; they can, however, succeed, fail, or be *etiolatic*. Successful ones are those performatives that actually do what they say. If someone says 'I forgive you' and they do not, then the performative has failed. These metrics are based solely in the speech act's *context*. For Austin, context is based on (1) a series of conventions made up of agreed procedures for the performative's enacting; (2) appropriate persons, words and circumstances around the speech act; and (3) an expectable, unsurprising effect once the performative has done its task. If these contextual conditions are met, it has succeeded and Austin (1975: 25) calls it 'felicitous'. Infelicity is the situation where the performative is 'unhappy' or fails. Etiolic speech acts, by contrast, are those performatives that neither succeed nor fail but where meaning is produced in a context where language is being used in a very self-conscious way (as in a poem, or a soliloquy or a pun). The etiolatic performative rests on the performer's ability to use language *as language*. In Austin's own words,

> Language in such circumstances is in special ways – intelligibly – used not seriously, but used in ways *parasitic* upon its normal use – ways which fall under the doctrine of the etiolations of language. All this we are excluding from consideration. Our performative utterances, felicitous or not, are to be understood as issued in ordinary circumstances.
>
> (Austin 1975: 22, emphasis original)

It is the 'All of this we are excluding. . .' that Derrida (1972) homes in on, because etiolatic performatives are such a pharmakon, a sign that inhabits both sides of an either/or dualism (Derrida 1972). Etiolatic speech acts are both performatives and not performatives at the same time. Difficult to categorise in Austin's happy/unhappy dualism, he simply ignores them. Derrida, however, does not. He notes that etiolations of language are excluded by Austin on the assumption that they were never intended to succeed. Thus serious speech acts are grounded in a metaphysics of presence, affirmed by the intentions of the actor and the sustainability of his/her context. Non-serious language is not. It simply quotes, repeats and reuses the serious original.

> In other words, does the generality of the risk admitted by Austin *surround* language like a kind of *ditch*, a place of external perdition into which locution might never venture, that it might avoid by remaining at home, in itself, sheltered by its essence or *telos*.
>
> (Derrida 1972: 325, emphases original)

*Contra* Austin, Derrida provocatively suggests that etiolatic language is not the aberrant exception by the rule of communication. He stresses the capacity of written language to provide meaning not through happy performatives' metaphysics of presence, but through etiolatic performatives' absences. Writing marks the iterative component of the performative – its sense of repetition that is slightly different each time. Iterability, however, means that immediate, situational context cannot be the final foundation of performativity's happiness. 'Repeatability implies repetition *elsewhere*' (Derrida 1972: 83, emphasis mine). Citations and grafts are possible, Derrida points out. Thus signs (be they speech acts or written language) can be outside a context. So while on the one hand iterability is the 'ditch' that language risks falling into (no stable bedrock of meaning), it is also the condition for its possibility (if we couldn't move language around, it would never communicate). The upshot of all this is that the

30

performative's contexts (procedures for inaction, surrounding words and expected result) are not denied, but they certainly cannot be the stable foundation for meaning in which Austin placed his faith.

Butler draws on both Austin's and Derrida's critiques to fashion her concepts of gender and sexuality. From Austin, she specifies the iterative, repetitive and 'doing' qualities of performatives (Butler 1993b: 231; 1990: 86; Jagose 1996: 85). From Derrida she takes the point that performatives can only seem to succeed or fail in a context of normalcy, but that normative context has no foundation, essence or core. The normality ensures a context's ability to under-write a performative's capacity to 'do something' but that normality must stem from the repetitive, iterative and allegorical nature of performatives themselves. There is a constant reference to an elsewhere, an 'outside' as Derrida (1972: 325) calls it. So performatives enact things by virtue of their constantly being done *elsewhere*, and that repetitiveness reflects and reinforces a normalcy to the performative.[6] Here she adds Foucaultian notions of power to the performative. It is an exercise of power, the power of discourse to define the normal, the typical, the appropriate. The point I would stress, though, is that immediate geographical context is decidedly dismissed as theoretically insignificant and uninteresting.

Adopting the poststructuralist point that everything is textual (or nothing is prediscursive) allows her move from speech to gender/sexuality. This brilliant move allows her to stake a number of important theoretical descriptions and insights. Foremost, of course, it provides her with a way to describe and discuss gender that is not essentialising or simply reducible to some immediate social context. Second, the move allows her to link the verbs of 'being' and 'doing' together. In a sense, her work is a sort of structuration theory for literary critics.[7] In the notion of performativity, social structure and human agency are mutually constituted, and their recursivity can produce unintended consequences (hence the iterativness). This point, in turn, allows us to see social action as moments or instances of broader power relations, but not simply explainable by them in some sort of 'last instance' because of the superstructures of (say) patriarchy and heterosexism. Ultimately they have no foundational base except for their own iterativeness.

In sum, we can see how Butler's ideas have solved a recurrent scholarly dilemma over how we understand that, despite individuals' resistance to power structures like gender, class, or sexuality, they remain constantly oppressive fac-tors in society. But she goes one better by suggesting a way of talking about these structures in an anti-foundational way, which emphasises just how socially con-structed and discursive these powers are. Moreover, she provides an interesting bridge between literary theory's emphasis on language and texts and social sci-ence's ken of social action. Yet in order to offer us all these gifts, her notion of performativity stresses the 'elsewhereness' of structure; immediate context and social action are only salient as allusions in a broader tale of power. This denial of context is currently being addressed by geographers interested in performativity (Rose 1996; Nelson 1999). Furthermore, as I argue in the following section, their arguments allow us a way to share insights between the closet metaphor and the performativities of gender and sexuality.

## Ignored geographies in performativity

Geographers have been excited and exercised by performativity recently (Pratt 2000; Gregson and Rose 2000; Nelson 1999). In both feminist and queer geographies, conceptualisations of gender and sexuality have been increasingly derived from Butler's work. For instance, Bell and Valentine (1995a) introduce their path-breaking collection *Mapping Desire* by staking sexuality as a performative. There and elsewhere, geographers have taken gendered and sexual performativity as axiomatic (e.g. Cream 1995a, b; Johnston, 1996, Rose, 1996; Bell and Valentine 1995b). From their spatial ken geographers are surreptitiously challenging Butler's denial of context. Interestingly, though, they have not made this point as a conscious critique. Nevertheless, by operationalising 'context' as location or place geographers have argued that it plays quite a central role in guaranteeing the reproduction of patriarchy and heteronormativity. McDowell's (1995) fascinating study of gendered bodies in merchant banks in the City, for example, shows how important the workplace *itself* is to perpetuating a culture where certain performances or comportments are rewarded (the male gym-body, for example) while others are punished (for example, when an *out* gay man is taped to his chair and sent in an elevator up to his superiors as a 'laff'). In a rather different way, geographers have emphasised the power of spatial context by conversely discussing how certain performatives are spatially transgressive and thereby showing the very social constructedness of gender and sexuality while politically reflecting on a human agent's performances as a means of resistance too. Bell *et al.*'s (1994) widely cited dialogic essay, for example, debates the transgressive potential performatives such as the gay skinhead and the lipstick lesbian in public, heterosexualised spaces. To their minds, performativity's contexts are geographic locations or situations, rather than speech acts or audiences. More to the point, however, this context is very much a part of that which is resisted through transgressive performances. It is therefore powerful, rather than a mere empty, inert stage of performance. Here Cresswell's (1996) notion of being 'out of place' comes to mind. And while Bell and Valentine certainly do not agree amongst themselves on the success or efficacy of sexualised performatives, they nevertheless agree that performatives cannot be understood without an appreciation of the spatialities in context. As they (1995a: 19) put it:

> By adapting Butler's (1990: 79) discussion of 'subversive bodily acts' to think about *subversive spatial acts* we can see how even the kiss of two men on the night bus home [or the sashaying of a man on an afternoon bus] can fracture and rupture a previously seamless (we might ironically say *homogeneous*) space [emphases original].

Two very related points are worth making. First, geographers have a rather different understanding of the relationship between social process and space from that of Butler, which allows them to stress context in a way that she simply refuses to do. Here I would note that geographers have stressed the spatiality of all social relations in a way that queer theory cannot. Pile (1996: 75), for instance, claims that performativity must be both situated practices and power-laden regulations. Johnston (1996) likewise assumes there is a mutual constitution of bodies and spaces in performativity. For queer theory, however, performance's space is naturalised into either a receptive audience or a stage metaphor. Where would we expect performances to take place, after all? On a stage (see Rozik 1994; States 1996). We can see this kind of troubling geographical imagination throughout performativity and queer theory. Jagose (1996), for instance, repeatedly refers to performativity's 'staged' quality or its 'staginess'. Likewise Parker and Sedgwick's (1995) edited collection metonymically foregrounds the stage-as-geography assumption by exploring performativity through performance and theatrical studies. They themselves explicitly identify actors and audiences as being key components to performativity. Patton (1995) as well argues promisingly that there has been a lack of interest in the spatial context of performatives, but retains this theatre trope by conceptualising context as a stage: 'there has been I believe an overemphasis on the actant-subject and a relative lack of consideration of the stage or context or field of the performance or performative act' (1995: 181).

The problem geographers have with the stage metaphor probably needs some unpacking here. When I was a student in the mid-1980s human geography was going through a sustained interaction with social theory generally. Geographers were not simply interested in importing social theory into the discipline. They were also emphatic that other disciplines had to appreciate the significance of a geographic perspective to their own theorising (Johnston 1997). And this was beginning to happen at that time (e.g. Giddens 1984). So I fondly remember a persistent refrain by my professors and in my readings that went something like: *'Space is not simply a stage or a container on which social processes play themselves out. Social relations don't take place on the head of a pin. They always take place somewhere. So geography matters.'*[8]

The problem for geographers with the stage metaphor was that it allowed theorists to divorce social processes from spatial context as a legitimate – indeed preferable – theoretical move (Gregory and Urry 1985). It was not merely an ontological error – since it assumed that topics of social theory (like race, class, or gender) 'really' existed abstractly. It was also an epistemological error, since it presumed that the best way to know a social process was to work at making it as abstract as possible, unsoiled by the messiness of 'real' time or space (e.g. Saunders 1980). The consequence of these fallacies, geographers pointed out, was that it essentialised terms and concepts in social theory by underwriting searches for their key, omnipresent, ubiquitous essence.[9]

Furthermore, it is interesting to point out that certain strands of scholarship on the theatre have theorised the stage in ways that mimic geographers' arguments about spatiality. Rozik (1994), for example, argues that the distinction between actor and stage might be productively blurred. He argues that portions of the actors' bodies and movements in Ionesco's *The Chairs* provide important non-verbal elements of staging for the production. The stage, in other words, is not simply incidental to or disconnected from the play. This strikes me as very much the same sort of point geographers have been making about social theory's will to understand its object of enquiry abstractly.

So not only is it ironic that performativity's stage metaphor has already been criticised for its aspatiality, but geographers' emphasis on the immediate context of a performative is also a decided criticism of Butler's preference for social power to be derived from the 'elsewhere' implied by her key terms of iterative-ness, repetition and citation. As the arguments stands, we seem to be forced into a silly choice: does the power of performativity 'really' come from the here and now of the performance or the social power built up elsewhere that underwrites immediate performance? Are performatives powerful because they are done in certain locations or contexts, or because they refer to the normalcy implied by their having been done previously, elsewhere? Staking the latter argument, it would seem, one cannot envisage the closet's immediate geography, never mind its power. Envisaging the closet as a powerful structuration of performance and performativity, however, can.

The second point I would highlight is that Butler is foremost a theorist and philosopher. Consequently, she seems to have considerable difficulty in using performativity in any sort of empirical way beyond the confines of a literary text or a 'real world' anecdote (Bordo 1992; Levenson 1998). For instance, Bell *et al.* (1994) specify the cultural context of performances that Butler's writings cannot accommodate, noting the variety of receptions drag might have for different audiences (cf. Parker and Sedgwick 1995). Here they echo a variety of scholars who find her work philosophically rigorous and enticing but exceptionally hard to utilise empirically around topics not immediately related to language or speech.[10]

It does seem as though Butler (1993b) has been especially vexed when her ideas are put to work in an empirical context, or perhaps more precisely from an empirical direction. For example, she has been sharply critical of her followers within queer theory who take drag to be the classic operationalisation of performativity. Butler (1990) originally cited drag as a very good example of how she thought performativity could be understood in queer contexts. Yet her later work chastises some of her most ardent followers, claiming that queer scholars have confused performativity with performance, the latter a mere insignificant step in a much broader and more significant if ethereal structure of power. Likewise, Sedgwick (1993) has stressed the dangers of using the concept too empirically, preferring to deploy it in the reading of literary texts rather than in real-space. In other words, to focus on drag *per se* is misguided: it is merely the outcome of a complex, highly abstract process (performativity) on which scholarship should focus.

Wrestling with the difficulties of appreciating both the real and unreal, the metaphorical and the material, without collapsing one into the other, geographers have done some fascinating work in 'taking Butler elsewhere'. Gregson and Rose (2000) compare the disparate empirical situations of car-boot sales and community arts programmes, suggesting that it might be helpful to think of space itself as a performative. In their own words:

> we want to argue too that it is not only social actors that are produced by power, but the spaces in which they perform. Contra to many geographical accounts discussed above, we maintain that performances do not take place in already existing locations: the City, the bank, the franchise restaurant, the straight street. These 'stages' do not pre-exist their performances, waiting in some sense to be mapped out by performances; rather, specific performances bring these spaces into being. And since these performances are themselves articulations of power, of particular subject positions, then we maintain that we need to think of spaces too as performative of power relations.

It seems to me that Gregson and Rose are struggling to find a way to appreciate the spatiality of social relations in such a way that does not privilege the 'elsewhereness' (or at the very least see its power as a spatial interaction between elsewhere and here – see note 6) Butler prefers nor the immediate context as an isolated, bounded stage.

From this review of how geographers have engaged with Butler's ideas I want to emphasise a few themes. The literary bias of performativity de-emphasises immediate, situational context in favour of the power of its 'elsewhereness'. This makes sense as a critique of the metaphysics of presence, but bristles over geographers' arguments about social theory over the past two decades. There also has been considerable difficulty in using ideas of performativity in topics other than speech acts. Both of these points have important implications for how we might understand the closet. They suggest that we might do well to notice the power of the closet as a spatial context of gender/sex performances. As well, we might simultaneously appreciate the linguistic dimension of the closet as a performative speech act. I am drawn to these points by considering how gay men have discussed the role of the closet in their own lives. They provide a helpful empirical venue to understand the performativity of the closet. The closet as materiality and metaphorical dimension in gay men's narratives is investigated in detail below.

## Speaking of the closet: personal narratives

I will stake the arguments developed above through a reading of six published editions of gay men's oral histories. Specifically, three are read from the United Kingdom: *Between the Acts* (Porter and Weeks 1991), *Proust, Cole Porter, Michelangelo, Marc Almond and Me* (National Lesbian and Gay Survey 1993), and *Walking after Midnight* (Hall Carpenter Archives 1989); and three from the United States: *Hometowns: Gay Men Write about Where They Belong* (Preston 1991), *A Member of the Family: Gay Men Write about Their Families* (Preston 1992), *Friends and Lovers: Gay Men Talk about the Families They Create* (Preston 1995). These texts allow us to explore the metaphorical and material dimensions of the closet at the scale of the individual subject.[11] They are recollections of gay life from a wide variety of times and spaces.[12] Informed by a geographic imagination that is sensitive to various ways in which the closet is being spatialised by these men, I undertake a textual analysis of their stories (Fairclough 1995). Perhaps most importantly, as Plummer (1995) has argued, one of the most important discernments of the self for gays is the coming-out narrative. The story of how we came out of the closet is a central point of reference for out gay men's lives.[13] It is a key, modern mode of self-identification, and coming out is an important topic spanning most of the texts.[14]

A series of practical concerns also obtain, making an unobtrusive tack in this kind of research seem especially apposite. It proved extremely difficult, and ultimately financially impossible, to devise an alternative research strategy to identify and interview a range of gay men (across class, ethnicity, age, HIV status, etc.) who actually inhabit the closet – especially internationally.[15] The difficulty in recruiting subjects in a diversity of locations in both the UK and the US meant that the edited collections were the most practical way of examining the closet at the scale of the individual. Because the research was on the closet *per se*, it seemed quite likely that obtrusive, dialogic research techniques would create a certain staged reality to the closet, which would be fundamentally at odds with Butler's approach. Likewise, the sensitive and private nature of the topic itself meant that even gay men who were ostensibly 'out' might not care to describe their closets for academic research. As a result, it was most appropriate to tap into already published discussions of gay men who could recall being inside the closet, or discuss times in their present lives when they were still positioned there. By using the oral histories to stake its argument, this chapter's point is not to claim that real-life individuals provide some logocentric adjudicator for what has been a rather abstract set of debates. Instead, it is to try and move beyond what Levenson (1998) has called the anecdotal attitude towards the empirical in Butler's work. It is to show how the closet's materiality informs performances of sexuality, and how it has performative qualities of its own.

Two potent themes emerge from these narratives that allow us to see the interrelatedness of metaphoric and material forces of the closet. First, these men make an interesting case for seeing the closet – à la Butler's heritage – as a linguistic performative. Its capacity to act through speech, however, is interestingly complex: for a number of men, coming out of the closet involved telling or saying their homosexuality. Likewise, these men tell tales of where not saying, not telling was essentially a performance of the closet. Second, the narratives also claim the closet as a material space through which their performances of gender and sexuality are done. The closet is often described as a specific place, and the coming-out process entails a physical move or migration from one place to another. The stories these men tell help us to see both the material and metaphoric forces at work through the closet.

### The closet as performative speech act

#### When saying is coming out of the closet

If we recall that the linguistic definition of a performative is when speakers do something by virtue of their speech, the possibility that the closet metaphor is linguistically related to performativity is made clear. For example, is having a sexual encounter with a person of the same sex the defining criterion of coming out of the closet, or is it actually stating, 'I am gay'?[16] If the latter, then the closet is resisted actually through a performative speech act. In other words, gay men break down the closet by telling or 'saying' their sexuality. Perhaps the most familiar example of the closet's relationship to linguistic performativity is when lesbians and gay men use speech to reveal their sexuality to themselves, to friends and family, to strangers. 'I am gay' is, of course, a performative. By uttering those words, one comes out of the closet. The context of those speech acts, of course, can be highly variable, and not always under the best circumstances for the speaker. Nevertheless, the locutionary force of this performative cannot be denied. Certainly it is one of the most common themes identified in the narratives, since most men discussed situations in their lives when they had to tell others that they were gay. The power of this voice destabilises the force of the closet. Plummer puts it this way:

> The narrative tells of a form of suffering that previously had to be endured in silence or may indeed not even have been recognised at all. The stories always tap initially into a secret world of suffering. They proceed to show the speaker moving out of this world of shadows, secrecy and silence . . . into a world which is more positive, public and supportive.
>
> (1995: 50)

And these acute events of speaking one's gayness signal just how powerful the metaphoric force of the closet is. The narratives collectively demonstrate the strength and courage gay men muster in order to 'come out of the closet' verbally – often (thought not exclusively) to loved ones. Read the excerpt below in which Christopher Wittke links the performative of coming out with the closet metaphor in his recollection about coming out to his sister Connie:

> . . . although I knew I would soon be telling all of them that the baby of the family was queer, I chose Connie for the season opener. I could then continue in ascending chronological order, telling my brother, my oldest sister, and mom. At some point after that, I imagined, we could all get together and figure out what to do about Dad, if anything.
>
> Connie was the first stop on the road from the closet and (I convinced myself) she was destined to be the most supportive.[17]

As it turned out, Connie's reply to him is 'Don't tell mom.' Her silencing maintains his suffering in a closet *vis-à-vis* his mother, which takes Christopher years to exit. In a quite different way, Trevor Thomas' coming-out tale to his family shows that performativity can work to break down the closet door from *both* sides:

> I'd been gay in America, quite actively so, and I'd come back and I'd got these boys [his sons], I decided, okay, I'm in London, I've got to devote myself to them. I got some weird idea at the back of my mind that I ought to let them know about my homosexuality. But found that my wife had told them! They kept hiding from me the fact that they knew, and I kept hiding from the fact that I didn't want them to know. I gradually came out one time and I just had to say, well all right now, I'll tell you the whole truth. You sit down and you listen.[18]

38

Here, the performative power of the closet is exercised by both those who are inside and those who are outside. Trevor and his sons needlessly work to maintain his closet from both sides of its door.

The significance of the coming-out performative is highlighted below also by Eric Latzky, who recalls how he and his close friend Matthew came out to each other in their sophomore year of high school.

> On an early winter day, the first revelation, Matthew to me: 'I have to tell you something. I'm . . . I'm bisexual.' Two days later, me to Matthew: 'I have to tell *you* something. I'm . . . I . . . I think I'm the same.' One week later, Matthew to me: 'I wasn't being completely honest. [Blurts this out] I'm gay.' Maybe another week goes by, me to Matthew: 'I wasn't being completely honest either. [Blurts this out] I'm gay too.'[19]

The halting speech, the time span, the half-truths, and the 'blurting' underscore not merely the difficulty and dangers of coming out, but also its power as a performative speech act.

In a rather similar context, coming out is linked directly with lust, desire and sex. For Tony, admitting he was gay at 15 to another boy was the key that opened the door for his first sexual experience:

> We lay there chatting and listening to the radio, then he said to me, 'Are you gay?' For a second I wasn't sure what to say. All sorts of things flashed through my mind. What if I admit it, what will they say at school? I didn't want to lose a good friend. In the end I decided to admit it in a joking sense; that way if the worst came to worst I could say that I was only kidding. 'Of course,' I said, nervously waiting for an answer. 'Then fuck me,' he said.[20]

In this example, the act of having sex with another boy seems to be less relevant to coming out of the closet for Tony than declaring his desire verbally. 'Doing by saying' seems rather more significant than 'doing by doing'.

The context in which this linguistic performative is done, it must be recalled, is not always on gay men's own terms, or to their benefit. Consider the 'forced' outing Norman faces in the 1940s when he was questioned by the police:

> Another time I went away down to Devon and Cornwall for a holiday. Rather stupidly I picked up someone and, although there was nothing to it, I took him to the hotel where I was staying. Well, this chap reported me, so when I got back to London, in my boarding house, Scotland Yard

came to see me. I was terribly upset and he said you've been reported as picking up soldiers in the park. I said, yes. I'm homosexual, I said. I tried to hide some pictures of Wolf. But, thank God, nothing happened. I was very worried about it.[21]

Although the audience might vary, and the context and consequences certainly do, each of these men exemplifies how coming out of the closet is a performative speech act. Speaking one's gayness, declaring one's identity is equated with a metaphoric mobility from inside to outside the closet.

### When silence is doing the closet

If a performative speech act is 'doing by saying', can its inverse also have performative force? In other words, can one 'do by not saying'? If so, then the performative force of the closet can be read from the variety of times the men's *silences* about their sexuality sustained their concealment or denial, and others' ignorance. Again, the complexity is fascinating. For example, David recalls growing up in England in the second decade of the 1900s. He performs the closet by not discussing or naming his sexuality for the simple reason that he does not have the word, the sign for it. In the following excerpt, he recalls his life at a boarding school when he was in his early teens. Without the ability to speak his sexuality, he remains closeted:

At that time I didn't regard myself as homosexual, I never thought of this word, nobody knew such a word. It was just something that you did. When I look back and think about my wasted youth, all the sex that you might have had and didn't have, it really wasn't true. This reminds me of Pushkin's famous remark, 'we are given youth in vain.' Rather well put, I think.[22]

Barry tells a similar tale of growing up as a British expatriate in Greece during the Second World War. Even though he finds ample opportunities to have sex with men, he still feels trapped in the closet, which he materialises quite explicitly in terms of not being able to tell his parents:

I can't say I was happy with my homosexuality, but I certainly didn't want to go and tell anybody and ask for help or anything of that sort. The thing that mainly worried me was the fact that my parents wanted me to get married, and that I could not, to use the modern expression, 'come out'. That was out of the question. My mother wouldn't have known what I was talking about. There's no doubt of that at all. And it would have been a cruelty to try and make her understand. And that is something that I am very prepared to put over to any of these young people who censure me for not being so willing as they are to come out.[23]

Silence on the topic of one's own sexuality does not exhaust the performativity of silence. For Mark Thompson, growing up in the small town of Carmel, California means he readily knows about the closeted adventures of his allegedly straight neighbours. Through his silence about their homosexual acts he hopes he can keep his own desires closeted:

> These currents of innuendo and gossip, which circulated around the town like a malicious ether, became known to me. They penetrated my frightened consciousness. I was afraid – achingly so – that someday such things might be true about me. So I kept the fact that I knew these stories closely guarded. It was one way, I must have thought, to buy protection against exposure of myself – as if knowledge, in this case, was power. Still, the hoarding of such secrets meant that I was living a life as lonely and prepossessed as the older gentlemen around me.[24]

One might insist that a performative *must* be a spoken utterance as Austin originally envisioned it to be. In this way, the silences that these men perform could never be considered performatives of sexuality. I think, however, this can be challenged on two grounds. First, if we treat speech–silence as a dualism that structures such a premise, we can begin to deconstruct the opposition between the two, seeing how each is predicated on the other. There is always silence when something is being said, any good deconstructionist would likely argue. Surely the force of not saying can also interpellate structures of gender and patriarchy. Second, by drawing on Derrida's (1972; 1982) hefty critique of speech's metaphysics of presence, Butler herself opens up the possibility that silence in a speech act might also claim a performative force. Silence must always already be partially constitutive of speech in that dualism. Moreover, Sedgwick's 'knowing by not knowing' epistemology cannot help but underscore the centrality of silence and omissions to the theory of the closet.

### When saying is doing the closet

But conceptualising the closet as a performative itself can show its more intricate, complex manifestations in the doing of gender and sexuality. For example, an interesting theme emerged in the histories where saying was performing the closet. As Eric Latzky's quote hinted above, the content of the speech in the speech act is integral to the power of its presence as a speech act. If we view the closet as externally imposed by the agency/speech of another, the force of scorn and ridicule is probably the most visible and harsh example of the way a speech act can perform the closet on another. A poignant example of this instance is recalled by Clifford Chase as he describes a fight with his brother (who later also turns out to be gay) over doing their chores:

> 'Stupid little faggot!'
>
> He had never called me that before. Or if he had, somehow it had never hit me in quite the same way. We started at each other a moment, and I think my face must have changed its shape. Perhaps I screamed. What I felt, and could not find words for, was this:
>
> *Not you, too.*
>
> I tore myself from his grasp and ran down the hall, an incredible and shameful grief pushing up behind my eyes. It was one of the last times until adulthood that I would really cry. Once in my room with the door closed behind me, that privacy did not seem enough either, and, as if to confirm the power of a future metaphor, I ran and shut myself in the closet.[25]

Ken closets Clifford with his hate speech, who in turn *literally* closets himself.

Alternatively, if we understand the closet to be a self-imposed structure, typically this power/knowledge exercise of the closet took the form of telling lies and misinforming others. By lying about his identity or sexuality, a gay man is concealing his sexuality through a speech act. Michael Bronski, for example, voices his frustration with living in Cambridge, Massachusetts, in the 1970s and 1980s because he continually encounters men who are closeted:

> Besides problems of finding easy places in which to pick up men, the problem with much of Cambridge – especially at Harvard – was the uptightness and the closetedness of so many of the men. For every encounter I had with someone like Ian – in which sex might be casual and unencumbered by emotional traumas – there were at least ten that were full of anxiety and fear. Many times the men I would meet in Harvard bathrooms refused to tell me their names or gave obviously false ones or lied about where they lived or attended school. Furtive was clearly the tone, which did not enhance the sex; the situation was frustrating and politically nettling.[26]

By lying, these men were inevitably using the closet as a performative speech act. By lying they were 'doing' the closet.

The closet as a performative speech could also take a more positive, productive form. Consider, for instance, the fascinating recollection of John Alcock, who lived in London in the 1950s. Here he recounts a specific kind of parole that was used by gay men in public space:

Parlyaree:    We created words for our own use and it came in from the ships into the dockside pubs and then the gay pubs. This was the language we spoke. It was great fun, of course, and it also enabled us to communicate with one another without other people being able to understand what we were talking about . . . 'Lallies' means legs; 'bats' – feet; 'Matinis' – hands; 'ogles' – eyes; 'ogale fakes' – eye glasses. Ears – 'Aunt Nell', which means hear as well. If you say 'Aunt Nell', you're saying to someone that you want them to listen to you. Then of course there's 'drag', which is quite universal . . . Quite a lot of people use it, especially in the theater. 'Camp' and 'butch' are both used by everyone now, of course.[27]

For John, saying is a form of manifesting the closet in the sense of keeping same-sex desire secret, hidden and concealed from straight men in the bars. Simultaneously, of course, we could also read Parlyaree as a means of breaking down the closet that isolated gay men from one another, and prohibited them from articulating and acting upon their desires. This use of secret language to navigate the closet in public space has even earlier references. Roy, for example, recalls life in 1920s London in a manner that resignifies the adverb 'so' into an adjective for 'gay':

We were 'so'. Have you ever heard that word? We were so. Is he so? Oh yes. He's so, and TBH [to be had] was a very famous expression. The sentence would go simply like this, well he's not really so, but he's TBH. And you would know exactly. I don't know when I first heard the word homosexual. I always remember somebody saying, 'oh she's a lesbian' and I didn't know what that meant. I didn't think women did that thing![28]

From these quotes, we can again see the way in which the closet takes the form of a speech act, where saying is performing one's sexuality in a very secret, concealed way. Here, saying is doing the closet not so much because of homophobia, but in spite of, or as a resistance to it. Being out of place, these gay men subvert the heteronormativity of highly masculinised public space in a way that leaves their objects of desire none the wiser.

### *When silence is breaking down the closet*

Silence can also be read in these texts as a strategy for breaking down the closet, though this theme was admittedly rare. Perhaps the most convincing way this theme could be traced through the narratives is exemplified by Alan Bell's account of how a colleague at work, Alex, came out. There is never a speech act, rather coming out of the closet is performed through bodily comportment and restyling (in much the same way Butler uses drag to exemplify what she means by performativity). While Alan and another gay colleague discuss ways to create a supportive situation in which Alex could tell them, Alex's corporeal 'doing' trumps 'saying':

> As it turned out we never did host a coming-out party for Alex. Inertia had set in, and when the notion resurfaced we realized there was no longer any need. Alex was now thoroughly out. Long sideburns framed his face and a single ring pierced his left ear. He'd thrown away all his old clothes and replaced them with black boots and wide belts, blue jeans and work shirts.[29]

Recognising the closet as a linguistic performative deepens and extends Butler's anti-essentialist arguments about gender and sexuality by focusing on the power of the speech act. Declaring one's sexuality is a means of coming out and this is a linguistic performative in the classic sense. Likewise, by remaining silent, by not telling one's sexual story, that which is known to the self remains unknown to others: heteronormative power is exercised once again. Conversely – though no less importantly – these men's narratives show how speaking through lies, secret languages, or the hate speech of others can perform the closet quite effectively. We might also begin to think about the ways in which silence could also be a strategy for assaulting the closet. Both simple and complex relations between performativity, speech and the closet are evinced in the narrative discussed above. There are, however, fewer linguistic and more material manifestations of the closet located in gay men's lives.

## The closet as a location

### *Sites for doing gender and sexuality*

While this chapter so far has examined the closet's performativity as a metaphor, we can also see how these men materialised the closet as a series of 'real' spaces they've encountered through their life course. These closets have different sizes and shapes to be sure, but they all express the power/knowledge of concealment, denial and ignorance of their sexuality in ways that emphasise their spatiality. It is part of the terrain on which they perform their gendered and sexualised identities. By either conforming to or transgressing norms of patriarchy or heteronormativity, the closet space is central rather than incidental to their performativity. It shapes their (in)actions. This was one of the most consistent themes to emerge from the narratives.

At some times, the closet seems to shrink to become the space of the body itself. Through deeply ingrained homophobia, several men talk about constantly disciplining their bodies, actions, speech and affectations in order to 'pass' for straight. They fear that their comportment will betray them and open the closet door. The intense isolation and alienation closeted gay men describe also materialises the body as a closet space. For Don, the closet is located inside his own sense of self: 'Within myself I treasured my secret.'[30] Others talk about becoming severely introverted, turning in upon themselves in order to avoid 'outside contact' that might open the door. Consider a passage from Philip, who recalls his closeted youth in Wakefield, Massachusetts, in this way:

> I realize that my awareness of myself, particularly of my sexual identity, was subsumed into a safe little suburban world. I was living in a cocoon, a cozy, warm, friendly cocoon that I had unconsciously spun out of all the pleasurable experiences that my town – and especially my high school – had to offer. It was a world of good grades and extracurricular involvements and C-block lunches with friends. If I was hiding amid all of this, I wasn't even aware of it myself. The truth is, it was easy to repress or ignore the occasional evidence of homophobia ('Don't wear a pink shirt on Thursday.'. . .) when otherwise life was so attractive.[31]

For Philip, the closet is not just manifest in his post-war suburban location. It can be found right down to the micro-scale of his own dress and comportment, which must be policed through a self-disciplining heteronormativity.

More common, perhaps, are narratives of the closet as a specific site in the city or town where these men lived. Urban space is examined more thoroughly in Chapter 3, but here we can begin to appreciate the hidden and secretive locations that enable and constrain these men's sexuality and desires. Typically the men discuss sites such as early gay bars, cruising areas, tearooms and cottages. These are closets where the gay body can act on its desire, where male bodies can sexually interact – but only so long as the site remains hidden or concealed. Recalling his youth in Charleston, SC, Harlan offers a complex, temporal geography of 'the Battery':

> It is the point of the peninsula where Charleston's two rivers, the Cooper

and the Ashley, meet to flow out past Fort Sumter to the sea, where all the streets eventually lead. . . . It was here we would walk after the Passover seders – where the moonlight disturbed me and later would not let me sleep. In the 1950s only white families would go there; in the seventies the hippies began to sit cross-legged under the trees. Go down there now on a Sunday afternoon, and you will see only blacks, coming from their neighbourhoods to take the air. . . . Old Charlestonians creep out to walk in the late afternoon, joggers dart by continuously. Underneath it all, like the shadows of the trees, is something else; it swells out, like the shadows do, in the evening. Come night and the Battery belongs to the gay of the city.

They all come down, sooner or later. Some circle hopelessly as moths around a beacon, never lighting. They never get out of their cars, their hysteria mounting with their fear. . . . They are all here: . . . the drag queens, the priests, and the society queens, scared teens taking their first steps, the daddies and good old boys in their pickups drunk and happy. And night lust licks around the edges of the chaste, charmed old city.[32]

Listen to how men describe their subject position being in the closet. It is not merely a metaphorical space (though it certainly is that). It is a way their physical geographies are demarcated. Tom's life in London during the 1950s and 1960s, for example, is spatially marked by closets that were bars, parks and cottages; whose secrecy heightens eroticism:

I spent my teenage years and early adulthood in putting on an act. I pretended to be normal, went out with and had sex with girls, while longing for my next secretive visit to the toilet in the park and Soho where I could have sexual pleasure with men. I recall when I first met gay men in bars and pubs it was marvelleous. Open, fun, outrageous — and a bit scary, but very exciting. . . . My main fears when in the closet were the fear and panic of being found out. . . . The sense of living on a knife's edge was a constant one.[33]

Likewise, Todd notes how in 1950s Baltimore, simply cruising in an automobile links subjectivity with geography, metaphor and material:

The cruising was mostly in cars. You stood on the corner and cars would circle around you. They'd drive by and slow down and that was it. You got in the car and went off and parked somewhere. The guys in the cars were quite often very closety types and probably chicken-hawks – men who like young kids – and then there were the types who liked young queens, the queenier the better. I was a very pretty kid, so I made out very well and was very popular.[34]

And Richard recalls the importance of the public washroom as a closet (a cottage) as an awakening of his same-sex desires:

> Cottaging is largely invisible to anyone not involved in it and it's quite incorrect to characterize it as a 'public nuisance'.
>
> Up to then I had come to associate cottaging with furtive, closeted homosexuality, with older or more conservative men, perhaps married, who sought to enjoy gay sex without facing the consequences of their sexuality. I considered such people as self-oppressed and cowardly, and it was a side of gay life that I distinguished sharply from my own. This attitude only changed when I came to consider cottaging as an *activity* rather than as a sexual lifestyle; it was something that anyone, myself included might do.[35]

In a rather different way, Malcolm notes how the closet manifests itself geographically at his work site:

> Now I am totally in the closet. This is really not through choice since I find it very difficult working with people and not telling them, but the nature of the work – residential child-care – makes it difficult since if the wrong people found out I could be out of a job. I feel absolutely no sexual interest towards the children under my care. . . . However, despite what I might say, the old suspicions about gays working with children will turn up, which would make life very difficult.[36]

For Malcolm, the closet is not a space of desire, it is a place that forces him to pretend. He draws on the normal script – confirmed so many times elsewhere – in order to be at his workplace (see also McDowell 1995).

In each of these narratives, the closet can be thought of as a space that is contextually important for the performance of sexuality – and the sex it drives. It is part of the iterativity of concealed sexuality since each space is slightly different, and thereby enables a slightly different manifestation of same-sex desire to be performed in each site. People are negotiating the possibilities of these geographies of desire and modify their behaviour accordingly. At a more complex level, Sedgwick's 'knowing by not knowing' is part and parcel of the complexity to the geographic imaginations of these men and the places they inhabit. They know when and where to go. So do 'straights' (to employ a fixed binary). But these places also retain a pretence of not being same-sex environments. Straight people –

and even many other sorts of queer people – would not necessarily have first-hand knowledge of these sites. They are known only vaguely: more unknown than known. And as an exercise of power/knowledge they maintain the marginality of homosexuality, even while offering the possibility that it could 'take place' somewhere. This, in fact, is developed further in Chapter 3, where the closet is spatialised at the urban scale.

*Migration and coming out of the closet*

If the importance of space to performativity can be detected statically in the position of 'being in the closet' or being in a closeted place, so too can it be read through mobility. This is the case with the men's accounts of 'coming out of the closet'. The expression's inherent mobility equates the subject's self-identification and truth telling to physical mobility of the body. A quite recurrent theme in these narratives was that of having to move to another place in order to know oneself as gay. It wasn't enough just to open the closet door; one had to leave its interior for a different location. John Preston makes this point rather poignantly in his introduction to *A Member of the Family*: 'I had to leave my family to be gay.'[37] Later he explains more fully that his need to leave home stemmed not so much from any hostility from his family, but rather their sheer ignorance of who or how he could be out of the closet and still with his family 'at home':

> I was infuriated by the frustration that overcame me when my parents couldn't talk to me about what was happening. I dismissed their reactions as bourgeois; there had to be a way out and they weren't going to help me.
>
> I decided to leave. I would go and discover my own life. One day I left a letter at home on my parents' bureau and announced my anger and my sense of defeat over what had gone on between us. I wasn't even going to tell them where I was going. I'd had enough. I expected they had as well.[38]

Cant (1997) has made this point more recently when he argues that lesbians and gay men so often lack a homeland upon which they can build identity. Because of the closet, their 'homelands' are likely places that have been harmful or difficult. To find a community through which they can be themselves, be 'out', they must migrate. Here the closet is a place that constrains freedom to act upon desire.[39] Barry voices this theme when he talks about his experience during the Second World War:

> I would have liked a relationship very much. Very much. And that's why I

went abroad, I got this job with the British Council to teach English abroad. I thought there would be more opportunities, that I would be freer. My first posting was in Greece. As it turned out it was Salonika, not Athens, and it was even more restricted than it had been in this country. But that was only the beginning because then I went to Athens and met quite a lot of people. And the whole thing became clearer. It became clearer what I should do in Greece to meet people.[40]

The existence of so-called 'Gay Meccas' like the Castro, or Brighton, or Greenwich Village provided a destiny that men spatialised as being out of the closet. The very spatial metaphor of 'Mecca' tropes on the powerful pull factors that certain places hold. These places are beacons for those who must remain closeted in their hometowns. As one New Jersey man described looking across the Hudson River:

> I knew nothing about my neighborhood [Greenwich Village] except that it had a reputation for being tolerant and artistic, perhaps a bit eccentric, so I presumed it would be a little easier to continue my carefully closeted gay life there than it had been in Newark, NJ where I'd been living with my father.[41]

Once the journey is complete, the self can be completed. The importance of the coming-out migration is noted below by Reed Woodhouse. He hints at how the place he came from could never be 'home', or familial space. Coming out meant moving out (in this case to Provincetown, Massachusetts – a largely gay town at the tip of Cape Cod). Moreover, the destination, rather than his origin, was home and family:

> What I eventually found in Provincetown was not safety but salvation, a place not to hide, but to shine forth. . . . What a relief it was to (eventually) realize that homosexuality didn't have to mean a lifetime sentence to respectability, that gay liberation could mean the freedom to *lower* your standards as well as raise them! Provincetown, friendly, tacky, and democratic, was exactly the right place to learn that lesson.[42]

Later he continues:

> Provincetown is the place where I've come closest to realizing a long-sought dream: a house of friends. This house, whatever its material form – guest house, apartment, or even the town itself – is a kind of second chance at a family, a second chance at 'home', a simulacrum (at

49

times) of love. In Provincetown I've had the chance to get 'family' right, and if not to go home again, to go there, happily, for the first time.[43]

In these stories, the spatiality of the closet is a quite powerful part of the performativity of gender and sexuality. For these men, resisting the hetero-patriarchal script does not just entail changing one's attitude, behaviour, dress, or style; it means having to relocate oneself, to leave 'home' and reconfigure it elsewhere. For some, coming out of the closet is spatialised as migration itself, for others coming out at home triggers a migration to a new place. It instigates the search for a new place to call home. These points fold into previous discussions of the closet as a metaphoric–linguistic space. Certainly these closets are multiple and manifold. They are found across a range of different locations, and their shapes have changed over time. Still, they are not just metaphors. They are spaces through which gay men perform their sexuality and gender (whether they con-form to or resist the norms). Either way, the power/knowledge of the closet is spatially important to performativity.

In both these manifestations (the closet as a stage for performativity, the closet as a performative itself), we can recognise, and therefore extend, Sedgwick's epistemology of the closet. For these men the closet was an open secret. They sought out gay-friendly (or at least gay-tolerant) locations that existed in the closet. These bars, cruising areas, cottages were open secrets. Sometimes they were secret places which only those in the know could find. Often they were closeted places that only became gay at certain times of the day. For those who had to move in order to 'come out', the process of migration entailed opening the very secret nature of their sexual desires. Without leaving the closet, sexual-ity could not be truthfully performed. Staying in one's hometown, or keeping one's job it seems, has meant a certainty of 'knowing by not knowing'.

## Conclusion

Returning to the vignette that opened this chapter, we can now better see both metaphoric and material dimensions to the closet at work on the bus. By speak-ing, the woman tries to closet the camp queen. By speaking, the camp queen comes out of the closet. By affirming the heteronormativity of public-bus-space, the woman attempts to closet the queen. By noting the immediate situation of the bus in a gay neighbourhood, the queen rejects the possibility of being closeted there; it is the homophobic woman who is out of place. The performativity of gender and sexuality is clearly at stake here. Yet it is instanciated in a perform-ance that derives its force not simply from 'elsewhere', but as a tarning of speech and space.

The purpose of this chapter has been to reassess Butler's notion of performativity with a geographic perspective on the closet. I have traced the concept's origins and shown the reasons for its considerable purchase in queer studies, especially problematising Butler's emphasis on the elsewhereness of performativity over its location and material performance. While there has been guarded caution about the meeting, geographers have been broadly supportive of Butler's notion, and it has allowed them to highlight issues of transgression potential in the relations between bodies and particular places. This work signals, by contrast, the importance of the place or 'stage' of the performance as part of the power of performativity.

In the light of this geographic perspective, I read a series of gay men's published oral histories in order for them to speak to the way we might begin to think about the closet's relation to performativity. The twining of the closet as both a metaphoric and material exercise of power/knowledge was emphasised. I argued that the closet might be thought of as a linguistic performative, a metaphoric space that is implicated in how we 'do' our sexuality. For many people, the classic template of being in the closet is remaining silent; the classic template for coming out is a telling. People's relations to a closet often involve a speech act. There is, however, no necessary relationship between speech and the closet, for inverse trends were also found. Saying could be a way for people to stay in the closet (simply by lying), and for those who are adamantly out, having to say or name their sexuality was seen as politically complicit with the enforced centrality and normalcy of heterosexuality. They refused to say they were gay because to do so would be to place them on the margins. Concomitantly, the closet was often a material space in which their bodies 'did' sexuality. These sites might be hidden bars, parks and beats, or even their own bodies. Likewise, coming out of the closet often entailed a physical migration away. Geographic mobility was a material expression of 'coming out'.

In many ways, this chapter has stayed very close to the methods of queer theory: examining books-as-texts, drawing on social as well as linguistic theory. In the following chapter, however, I treat the urban landscape as text, in order to see how the closet materialises in the city in ways that a leading scholar on the geographies of capitalism fails to recognise.

## Notes

1  Here I am also signalling Butler's (1997b) examination of performativity through hate speech.
2  Sedgwick (1993) specifically examines the performativity of shame in, it seems, producing the closet. She does this through a textual analysis of Henry James' *The Art of the Novel*.
3  See Jagose (1996) and Sedgwick (1993) on this connection for examples.
4  Bell and Valentine (1995b) claim that there should be no need to review performativity

for critical human geographers. I do think a review is necessary, however, for two reasons: first because it logically produces an internal critique of her work and second, it has been my experience that many geographers have a difficult time with Butler's often turgid and exclusionary prose.

5 Admittedly, however, these are structures in a poststructural sense: without fixed or prediscursive foundations.

6 As a geographer, I would see this 'elsewhereness' not disconnected from immediate context, but rather as a form of spatial interaction that is part of the power of the performative.

7 I realise this is not entirely correct. Structuration theory (Giddens 1984) is ultimately a foundational social theory in the sense that it presumes a more or less fixed, prediscursive set of social structures, not to mention a rather cogent, decidedly centred conception of subjectivity bounded in the agent. In its early formulation, moreover, it tended to emphasise social structures of class alone. Butler's performativity, by contrast, adamantly denies that there is anything that is prediscursive, and obviously is specifically geared towards structures of gender and sexuality. Nonetheless, I make the point that both are attempting to solve the perennial problem in social explanation and description of the interrelatedness of structure and agency. Making this point also helps me stress the significance of spatiality to social relations, since this was a key point in Giddens' overall framework (see Cloke, Philo and Sadler 1993).

8 For some examples of this sort of argument from that era see Massey 1984; Gregory and Urry 1985; Wolch and Dear 1988.

9 Politically, of course, it also meant that geographers would be for ever relegated to mere empirical description, while disciplines like sociology did the harder work of theorising about and with key terms and ideas. It is perhaps no coincidence that geographers' intervention into social theory came at a time after several departments in the US were closed or amalgamated within the social sciences.

10 Indeed, Butler's more recent book *Excitable Speech* (1997b), has her returning performativity back to its Austinian roots, in a sense. By focusing on hate speech or the US military's 'don't ask; don't tell' policy she seems to have reigned in performativity away from a theory of social process generally towards speech and language specifically.

11 In order to maintain the flow of the text, I use footnotes to cite the excerpts taken from the narratives. They are abbreviated as follows: *Between the Acts*: BTA; *Proust, Cole Porter, Michelangelo, Marc Almond and Me*: PCP; *Walking after Midnight*: WAM; *Hometowns*: HT; *A Member of the Family*: MOF; and *Friends and Lovers*: FAL.

12 The narratives recount lives of gay men who live in the UK and the US, and span the entire twentieth century. This range is significant in the light of Sedgwick's argument that the closet is a defining feature of this century.

13 The decision to use published oral histories, of course, hides a fascinating methodological dilemma. Given that these men have already told their story publicly, they are 'already' outside the closet. Would it not be better to interview people who are presently in the closet, to get a sense of its immediacy and interiority? The answer is, of course, methodologically yes; but ethically and practically, such a method would be so difficult that I found this the best, albeit not perfect, solution.

14 The common framework of coming-out stories include suffering, epiphany and transformation and Plummer (1995) sets these in the five standard modernist plots: taking

a journey, engaging in a contest, enduring suffering, pursuing consummation, establishing a home (based on Propp 1968).

15 The choice of the US and Britain also raises the postcolonial issue of the closet as an ultimately western position. This point is taken up explicitly in Chapter 5.

16 For debates over a reliable definition of homosexuality in social surveys see Michael *et al.* 1994.

17 Christopher Wittke, MOF, p. 275.

18 Trevor Thomas, BTA, p. 68.

19 Eric Latzky, FAL, p. 192.

20 Tony, PCP, pp. 8–9.

21 Norman, BTA, p. 32.

22 David, BTA, p. 42.

23 Barry, BTA, p. 131.

24 Mark Thompson, HT, p. 35.

25 Clifford Chase, MOF, pp. 160–1.

26 Michael Bronski, HT, p. 183.

27 John Alcock, WAM, pp. 51–2.

28 Roy, BTA, pp. 75–6.

29 Alan Bell, FAL, p. 266.

30 Don in PCP, p. 31.

31 Philip Gambone, HT, p. 88.

32 Harlan Greene, HT, p. 63.

33 Tom, PCP, pp. 155–6.

34 Todd Butler, WAM, p. 80.

35 Gareth, PCP, pp. 119–20.

36 Malcolm, PCP, p. 100.

37 John Preston, MOF, p. 2.

38 Ibid., p. 2.

39 For a discussion of the metonymic relation between closet and world see Brown 1996.

40 Barry, BTA, p. 128.

41 Arnie Kantrowitz, HT, p. 258.

42 Reed Woodhouse, HT, p. 222.

43 Ibid., p. 227.

# 3

# PRODUCING THE CLOSET IN URBAN SPACE
## Spatiality, sexuality and capitalism

### The sex capital of New Zealand

During an otherwise cold and bleak winter in 1996 the city of Christchurch, New Zealand, became embroiled in a series of controversies around sex and space in its inner city. Earlier that year, the city's daily newspaper discovered that Latimer Square (a public park on the eastern side of downtown) was the site of a thriving youth prostitution market, where homeless teenagers were selling sex to support drug habits (Keenan 1996). In August, a national news programme featured Christchurch as 'the sex capital of New Zealand', reporting that this modest-sized city of 288,000 boasted 22 massage parlours, 297 escorts and an estimated 370 sex workers ('Assignment' 25 July 1996; Harris 1996: 10). The programme suggested that the sex industry in the city was out of proportion to its size. Likewise in June a group of concerned citizens voiced anger at the number and concentration of massage parlours and strip joints in the central city which used degrading images of women's bodies to advertise their services in public space (see *The Christchurch Star*, 24 June 1996: 1; 9 August 1996: 1; *The Christchurch Mail*) A certain sensitivity had been building up around the city's salacious image. Popularly known as 'the most British city outside of Britain', Christchurch portrays itself as a small quiet, conservative, middle-class white settlement on the edge of the empire. Its popular landscape hearkens back to a simulacrum of the asexual, upper-class districts of Victorian England and earlier, with manicured public parks, neo-gothic architecture and British toponyms (see Figures 3.1, 3.2, 3.3).[1] However incongruous and uncharacteristic they may seem, issues of sexuality and space seem very much on the local agenda these days in the city.

The inner city in Christchurch prompts a number of interesting theoretical issues around the materiality of the closet in urban space. For instance, we might find it a useful case study to explore Lefebvre's notion of contradictory space – how does the city spatially reconcile its 'Garden City' image with its growing

*Figure 3.1* Christchurch Cathedral. Located at the very centre of the inner city, the cathedral symbolises a certain chaste, British-colonial heritage. This was a city founded partially on religion. (Photo by author)

*Figure 3.2* 'The most British city outside of Britain', the inner city of Christchurch is
noted for its neo-gothic colonial architecture. Pictured here are the former
Canterbury Provincial Buildings on the banks of the Avon River. (Photo by
author)

reputation as 'Sleaze City'? More interesting, however, are broader questions on
relations between sexuality and space that might be teased out of that landscape
by considering the closet as spatial exercise of power/knowledge. In this chapter
I show how the closet can be an important material dimension to the relations
between sexuality generally (and gay men specifically) and the city. By employing
a multifaceted textual analysis of the Christchurch landscape, I will show how the
closet is more than just a spatial metaphor at the urban scale. It is a material
production of heterosexism and is inscribed in urban space. The closet also
enables gay desire to be commodified for profit. And in that way the closet is a
production not only of heteronormativity in urban space, but simultaneously of
capitalist relations.

*Figure 3.3* Body in the landscape. Not merely British imperialism but Victorian culture is signified in the landscape here with a statue of Queen Victoria in Victoria Square in the northeastern part of inner-city Christchurch. (Photo by author)

Theoretically, this reading draws on, and speaks to the work of, French urban theorist Henri Lefebvre (1991; 1996). There are several reasons why I engage with Lefebvre's writings *vis-à-vis* the closet. Lefebvre was especially interested in understanding the fundamental spatiality of all social processes. He was especially suspicious about using only vision to detect the production of space. Since the closet metaphor often turns on a visual trope (its contents being unseen), his work seems especially apposite. Lefebvre certainly did not ignore the role that sexuality played in the production of space in capitalist society. In spite of the numerous references to sexuality in his writing, Lefebvre has surprisingly little to say about the relationship between different forms of sexuality and urban space (Blum and Nast 1996). His comments on the topic are rather disparate and directed towards more central themes on the types of space at work in the city. Tempting as those snippets may be, they do not add up to a robust theory of sexual relations and urban space. His remarks on commodification through visual fragmentation of the body, consequently, must be made more sensitive to multiple sexualities. Most troubling is the fact that, like so many authors of his time, Lefebvre seems to operationalise the term 'sexuality' in heterosexual parameters exclusively. Sexuality, it seems, is a productive relation, but one poled by different genders, which only ever desire their other.[2] How might Lefebvre's thinking about urban space be enhanced with a sensitivity to different sexualities? The brief account of Christchurch above, for instance, hints at a quite explicit and visual production of heterosexual spaces in the city. Yet in exactly the same space, the city's gay nightclubs, saunas and cruise clubs are also located, albeit largely invisible to the public eye. Putting it all too crudely, do these different sexualities produce urban space differently? Lefebvre's work has been instrumental in rethinking the relations between society and space for the past thirty years (Gregory 1994; Soja 1996). This widespread influence makes his heterosexism especially problematic. It has already begun to be addressed in the work of Blum and Nast (1996). They juxtapose readings of Lefebvre and Lacan in order to argue that Lefebvre's reading of alterity is principally located around heteronormative gender relations, and to call for ways to extend his thinking in queerer directions. This chapter tries to answer that call by reading the Christchurch landscapes through Lefebvre's writings.

I proceed in three steps in order to examine the role the closet plays in the production of urban space. In 'Sexuality and the production of space', Lefebvre's general theory of space is outlined, emphasizing his complicity with heteronormativity and his consideration of visual epistemologies in understanding spatial structures in the capitalist city. Following Duncan (1990), two readings of the Christchurch urban landscape are offered with the closet in mind. The first, following Lefebvre's suspicion of 'abstract space', tempers his critique by showing how official, bureaucratic-scientific representations of the city (through tourism literature and geographic information systems [GIS] mapping), while certainly not revealing a sexualised landscape on their own, do hint towards it.[3] The second approach is more obtrusive. Relying on interviews with key inform- ants, I pursue the points raised above by showing how the closet works as an important part of the landscape in the inner city. I draw on Knopp's (1992; 1995a) path-breaking work on the relations between a capitalist market, sexuality and space, as well as Harvey's (1996) notion of 'spatial fix' in his theory of capitalist accumulation. These authors call our attention to the relations between capitalism and heteronormativity, and to the inseparability of the urban landscape from capitalist forces, respectively. Through his appreciation of the materiality of the closet in Christchurch, Lefebvre's insights into the relations between capitalism, sexuality and the city are extended in more queer-friendly directions. I demonstrate how the closet materialises in urban space as a spatial fix in order for same-sex relations to be successfully (and spatially) commodified for profit.

## Sexuality and the production of space

Lefebvre's central theoretical contribution to geography has already been alluded to in earlier chapters. It is his insistence that space is no mere stage or container on/in which social processes unfold. Space is a necessary and fundamental elem- ent of all relationships. Consequently, all social scientists must recognise that social relations only exist through dimensions of time and space. They are there- fore always affected by their spatial qualities (that social relations affect space seems more easily appreciated (see Kearns 1992). Lefebvre's remarks on the relations between sexuality and the production of space, however, are somewhat more random and desultory. Still, there are a number of interesting themes that can be teased out of his couched remarks. At first, Lefebvre reproduces the ideology that sexuality is a productive relation to the extent that it is directly related to the re-production of the species (e.g. Lefebvre 1991: 32). Described in this way, sexual relations are largely relegated to the private sphere of residence and home, and in service to the capitalist mode of production to the extent that a labour supply is functionally maintained (Peet 1991). And yet later in his analysis, Lefebvre seems to realise how superficial, though nonetheless accurate, this read- ing is. His thoughts on sexuality and space become much subtler and less reduc- tionist.[4] For example, at several points in *The Production*, he raises the issue of the phallocentricity in the production of space, anticipating contemporary archi- tectural critiques (e.g. Betsky 1995). Most generally, however, Lefebvre is keen to point out a constant tension between abstract space and (true) sexuality. *Abstract space* is the space of modernity, produced by capitalism and the modern state, and is expressed through 'representations of space' (Lefebvre 1991: 53,

59

38–9). Blum and Nast (1996: 572) summarise it well in a historical context of modernity:

> Progressive abstraction of spatiality and labor began when capital and the state seized upon abstract Euclidean space to frame and situate the world. Euclidean space allowed three-dimensional social relations and private property to be projected 'scientifically' on two-dimensional surfaces. Lived forms were reduced to planar configurations that could be much more easily managed through written means, including directives, quotas, diagrams, and maps. In other words, by placing the world inside Euclidean space and then geometrically manipulating it, capital projected and managed the world through surfaces.

For Lefebvre abstract space conjoins the representations and realities not just of capitalists, but also of modern city planners, economists, bureaucrats and other such scientific urban experts. Ever hostile to the natural, the creative and the body, abstract space *denies* the sexual.

> A characteristic contradiction of abstract space consists in the fact that, although it *denies* the sensual and the sexual, its only immediate point of reference is its genitality: the family unit, the type of dwelling (apartment, bungalow, cottage, etc.), fatherhood, and motherhood, and the assumption that fertility and fulfillment are identical.
>
> (1991: 49–50, emphasis mine)

In the passage above, Lefebvre is trying to signal the obfuscatory role abstract space can have: it hides and conceals personal and intimate relations like sexuality. Abstract space is the spaces of authoritative rationality that have become so hegemonic in recent times. Abstract space has little room for desire. Thus, he argues that sexuality is often confined in artistic works and images, which he calls 'representational spaces' (Lefebvre 1991: 50).

In one of his more prescient and extensive passages on sexuality, Lefebvre shows how the body itself is a factor in the production of space. By producing gestures, traces and marks in space, the body is made socially visible in the landscape, and hence meaningful. He readily points out that in public spaces, only certain sorts of locations come to produce space through sexuality, like red-light districts at night. He is most interested in pointing out that these sexualised relations are also relations of capitalism, but not simply in a general, functional sense. They can become commodified themselves, through prostitution for instance. In other words, these spaces are produced not merely for the stimulation and satisfaction of desire, but also for potential profit. Here, the body –

especially the female body – is abstracted and shattered into fragments that are visually displayed where profit and sex collude:

> Confined by the abstraction of a space broken down into specialized locations, the body itself is pulverized. The body as represented by the images of advertising (where legs stand for stockings, the breasts for bras, the face for make-up etc.) serves to fragment desire and doom it to anxious frustration, to the non-satisfaction of local needs. In abstract space, and wherever its influence is felt, the demise of the body has a dual character, for it is at once symbolic and concrete: concrete, as a result of the aggression to which the body is subject; symbolic, on account of the fragmentation of the body's living unity. This is especially true of the female body, as transformed into exchange value, into a sign of the commodity and indeed into a commodity per se.
>
> (Lefebvre 1991: 310; see also pp. 166–7, 204–5)

The central theme here is Lefebvre's insistence that not only does sexuality produce space, but it does so publicly through capitalist social relations that commodify the product. Following Gibson-Graham (1997), it is not so much that a market for women's bodies, or sex generally, is fashioned though this kind of space; it is that the accumulation of capital (which is ultimately premised on the extraction of surplus value in production relations in society generally) always happens through space.

A second and related theme to note is Lefebvre's concern with visuality. Vision has a primacy for sexuality. Quite provocatively, he claims that we now come to know our sexual desires primarily through visual signs and regimes. Yet like many French intellectuals, Lefebvre was deeply suspicious of what Jay (1993) has called 'scopic regimes', the over-reliance on the visual as an epistemological guarantee.[5] The modern eye dominates all other senses. For instance, in his critique of abstract space, Lefebvre (1991: 287) notes an important element is its optical or visual format:

> In the course of the process whereby the visual gains the upper hand over the other senses, all impressions derived from taste, smell, touch, and even hearing first lose clarity, then fade away all together, leaving the field to line, color, and light. In this way a part of the object and what it offers comes to be taken for the whole. . . . Finally by assimilation, or perhaps by simulation, all of social life becomes the mere decipherment of messages by the eyes, the mere reading of texts. Any non-optical impression – a tactile one for example, or a muscular (rhythmic) one – is no longer anything more than a symbolic form of, or a transitional step towards, the visual.

He insists that we move away from visual understandings of space towards other sensory epistemologies, and he includes sex and sexuality among these alternatives. Lefebvre problematically seems to be suggesting that a 'true', 'real', or authentic sexuality (one not usurped by capitalist relations) would block the fragmentation of the body. The reasons behind his suspicion are that the visual can trick us into not seeing particular dimensions to social relations – especially in space.

> A further important aspect of spaces of this kind is their increasingly pronounced visual character. They are made with the visible in mind: the visibility of people and things, of spaces and of whatever is contained in them. The predominance of visualization (more important than 'spectacularization' which is in any case subsumed by it) serves to conceal repetitiveness. People look, and take sight, take seeing, for life itself. We build on the basis of papers and plans. We buy on the basis of images. Sight and seeing, which in the Western tradition once epitomized intelligibility, have turned into a trap: the means whereby, in social space, diversity may be simulated and a travesty of enlightenment and intelligibility ensconced under the sign of transparency.
>
> (Lefebvre 1991: 75–6)

What appears in space as simply 'just there', or 'there naturally', diverts our critical attention from the broader social forces that produced those patterns. The point I would stress from Lefebvre's work is to be suspicious of using only vision to appreciate the ways space can be produced.

So while his thinking romantically essentialises a pure, power-free sexuality and it does not acknowledge the possibility of alternative sexualities producing space, his work does suggest important lines of enquiry into that relationship. He alerts those of us who trade in abstract space to consider its power to conceal or deny – in other words to closet – the sexual production of urban space. Lefebvre has also highlighted the capacity of the body to inscribe urban space, especially as it becomes fragmented through commodification. This point has certainly grown into a current topic in social theory (Nast and Pile 1998). In addition, Lefebvre's remarks on the epistemological primacy of vision – and a healthy suspicion of it – suggest that seeing sex in urban space is fraught with difficulty. In what follows I use these insights to explore the inner city of Christchurch to see how far Lefebvre's insights get us in operationalising the closet as part of the sexual geographies of the city. For men who have sex with men, closets abound in the inner city. They are the secret, hidden and concealed – Lefebvre might call them invisible – places where they can enable their desire. There, the closet becomes an important spatial fix, a production of space(s) that functionally enable the commodification of sexualised relations to occur.

## Sexuality and abstract space

How might these comments be used to investigate the role of sexual relations in a particular locale? I will begin to explore this question by situating the closet in a more general consideration of the way sexuality is part and parcel of the production of urban space. In the case of Christchurch, three themes resonate with Lefebvre's comments: (1) the tension between abstract and sexual space, (2) the role that sexual relations have in producing urban space, and (3) the importance of the body's commodification and fragmentation in the production of sexualised space. First, we might consider just what sort of meaning of the inner city is produced through abstract representations of space. This tack has the dual advantages of establishing Lefebvre's point empirically, and also giving some context to what sort of place central Christchurch is. To assess critically Lefebvre's suspicions, two rather different texts that embody abstract space were read with an eye towards sexuality: planning and tourist literature, and locational data manipulated with a GIS map.[6]

The inner city district is culturally and administratively defined as the area due east of Hagley Park, bounded by a ring of four arterial avenues, and centred on the Christchurch Anglican Cathedral. Approximately 3 square kilometres in size, it houses about 1,000 retail properties, as well as the site of local government. Historically, the area has been seen as a place of work rather than residence, though there has always been a small number of central city residents (e.g. Ferguson 1994). More recently, with the extension of opening hours for licensed establishments, the area has become a hub of nightlife (Lawn 1995; Koeksen 1993; Canterbury Tourism Council 1995).[7] Not surprisingly, representations of space from social and tourist planners make little mention of the bustling sex industry located there. The abstract space produced through these official visions, as Lefebvre predicts, does not directly discuss sexuality's role in space. Compare the following representations, one from a planning journal, one from a tourist brochure, and one from the city's annual report:

> The City Mall . . . is now a success for everyone to share. And now Worcester Boulevard is a clear statement of rediscovering the quality of the city's layout. Suddenly, the puzzle seems to fit together: Cathedral Square as the most important public space, the Boulevard as the connection between the Square and Hagley Park, Art Centre, and Museum, Columbo Street as the city's retail 'lifeline', Victoria Square and the river Avon as symbols of both European and Maori respect for the natural environment; the Press Quarter behind the Cathedral as New Zealand's last remnant of European urban history, and last but not

least, the Anglican Cathedral itself as Christchurch's most important building.

(Koeksen 1993: 23)

Residents and visitors congregate in and around Cathedral Square and Victoria Square. From here, you can set out to admire the Victorian Neo-Gothic stone buildings riding in an antique tram, vintage car, horse-drawn carriage or river punt. The country's first casino is within walking distance from here, as are the major inner city hotels. This compact city demands to be taken in at a dignified pace; hire a bicycle or simply stroll along the river bank.

(Canterbury Tourism Council 1995, no page)

The council continues to emphasize upgrading the quality and liveliness of the central city. It must be a safe and attractive place where Christ-church residents and the growing number of overseas visitors can mix in a way which is both enjoyable and, through tourism, beneficial to the local economy and employment.

(Christchurch City Council 1996: 7)

In each representation, rational planning for social interaction and commerce are the discourses through which the central city is spatially produced. This point echoes through the findings of several geographers who make the point that urban images are very important in the capitalist accumulation process (Harvey 1989; Smith 1996). They also resonate through the newer literature on the importance of cultural production in urban economies under post-Fordism (Zukin 1996). Those representations above, however, do not – and in many ways could not – admit that these discourses also operate through commodified sexual relations, in part because they conflict so directly with the chaste and religious colonial images upon which Christchurch trades (Jacobs 1996). So here we can see the tension between abstract and sexual space. Of course the irony is that sexual relations may very well help make the central city such an 'enjoyable' site for 'mixing' certain tourists and residents. Perhaps for some people, part of what makes the central city so exciting and vibrant is its bustling sex trade. Maybe part of what attracts certain tourists and bolsters the local economy are the sex-on-site venues. Indeed, recent research has suggested that the sex tourism is worth $NZ 15 million to the New Zealand economy annually (*The Dominion*, 4 November 1996: 2). Sexuality may not directly appear in this production, but it can exist subtextually, or appear when abstract space is juxtaposed with other productions.[8] Furthermore, in the context of other texts, we can begin to read the text of abstract space differently, hinting at the economic role sex plays in the production of urban space.

A more deliberate attempt (using GIS and rudimentary spatial statistics) at visualising sexuality through abstract space also reveals the tensions, but nonetheless shows the spatial extent to which sexuality might define the inner city. In New Zealand, prostitution is not necessarily illegal, but everything that makes it possible is. It is illegal to keep a brothel, to live off the earnings of prostitution, and to solicit (see *New Zealand Statutes*, Crimes Act, 1961, sections 147, 148, 149, pp. 67–8). For these reasons massage parlours and saunas have historically been the venue for sex work in the country. These locations provide a spatial cover where the client is said to pay for a massage and workers must pay shift fees to be there as well. Parlours also provide a degree of safety, information exchange and community for the workers.[9] While there are some street workers in Christchurch, the majority of sex work in the city takes the form of the massage parlour or bathhouse. Plotting the locations of such public sex establishments revealed that all but one parlour can be found within the four avenues. As a production of abstract space, Figure 3.4 illustrates the general pattern, by plotting the spatial standard deviation and mean centres of sex-on-site venues.[10] The larger shaded circle is for straight massage parlours; the smaller one is for gay venues. This visualisation maps sexualised urban space as largely coterminous with the whole inner city of Christchurch.[11] At some level this map argues that urban space and commodified sexual space are equivalent.

Turning away from abstract spatial representations, we can appreciate Lefebvre's remarks on the commodified and fragmented body. Here Duncan's (1990) textual approach to landscape is more helpful insofar as it draws our attention to representational spaces of sex. Many, but by no means all, the parlours and strip bars in the city advertise their presence in the streets of central Christchurch with highly objectified images of women. Affirming Lefebvre's critique above, Figures 3.5 to 3.9 show the preponderance of objectified, heterosexualised female body-images in the inner city. Silhouettes of women's bodies, murals of large-breasted women or women's faces, eyes, mouths, etc. abound in the central city. Women are objectified as the word 'GIRLS!' is repeatedly found on signs, doors and, in one case, the entire side of a building. Through an explicitly visual regime, women's bodies are fragmented as they are commodified. Pieces of the body displayed in public space come to stand for the commodification of heterosexual relations.[12] From these heterosexist images we can see that sexualised relations do not just take place in space, but take place through it.

This discussion so far has drawn on and extended many insights in Lefebvre's work. The sexual production of urban space does take place, as central Christchurch has demonstrated. It can be hinted at through abstract space, though that methodology needs context from other representational strategies. Finally, sexual space is reproduced visually. The body is visually fragmented into pieces, as the massage parlour signs show. The fragments show how sexual relations are

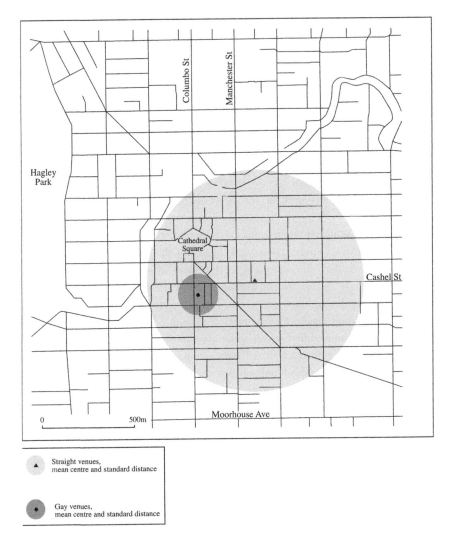

*Figure 3.4* Sexualised public space in Christchurch's inner city

commodified. Lefebvre's thinking seems especially appropriate to understanding the production of space in Christchurch. But where is the closet in all of this?

## Sexuality and space: capitalism and the closet

As Figure 3.4 hints, the Lefebvrian geography of Christchurch offered so far is incomplete. Quite simply, it belies the closet in the production of urban space. There are also gay bars and male sex-on-site venues that also comprise the sexual

*Figure 3.5* Wicked Willie's massage parlour. (Photo by author)

geography of the inner city.[13] We can see that they are located within the same space as the straight venues, albeit in a much smaller concentration.[14] The standard distance for gay venues has a mere 125 metre radius, while straight clubs have a radius of just over half a kilometre (623 m). This clustering within a cluster is partially due to the smaller number of queer venues overall in the city. Figure 3.4 also implies the operation of a sexual closet in the production of the central city.

This point is evinced in two ways. Figure 3.4 draws on and reproduces a closet in abstract space. The use of standard-distance circles was deliberate on my part. On the one hand, the larger circle-in-the-square image illustrates nicely how the inner city in Christchurch generally is a space of (hetero)sexuality. On the other, I could have just plotted each venue on the map. I rejected that option because gay life in Christchurch is so closeted (see below). There was something rather

*Figure 3.6* Blondie's massage parlour. (Photo by author)

unethical about 'outing' these venues on a map. (Indeed, images of the venues later on in this chapter have been altered to protect their anonymity, their closetedness.) The point here is that, by further abstracting spatial representations of sexuality, the closet has been reproduced in both simple and complex ways. Simply put, the locations of gay or same-sex venues remain occluded with greater abstraction. Yet more complexly, the map paradoxically reveals gay space as well.[15]

We might be tempted further to assume that the insights above can also describe the production of gay sexual space in the inner city. As discussed in Chapter 1, however, recent work in queer theory has stressed the importance of the closet in understanding the unique oppression of lesbians and gays. If there was no closet in the central city, why wouldn't opponents of the straight sex-on-site venues (from the introduction of this chapter) include the same-sex venues in their campaign – especially given the stigma homosexual acts still have in society?

*Figure 3.7* Alamein bathhouse. (Photo by author)

The two sorts of venues exist in exactly the same space of the city! During all the controversy around the sexual space in the inner city, *not once* has the presence of all-male venues been mentioned. Why has it gone so unseen – even beyond the spaces of representation? It follows that an understanding of the production of 'gay space' must be sensitive to the formation of the closet in urban space.

The significance of the closet, I would argue, can be threaded through Lefebvre's own discomfort with visual epistemologies. While Figure 3.4 implies a spatial similarity to straight and gay venues, it also suggests they only produce urban space differently in terms of number of venues, or different spatial ranges. The intense visual presence of straight venues draws our attention away from the way parlours normalise male heterosexuality. Gay venues, in stark contrast to

*Figure 3.8* Cat Ballou's bathhouse. (Photo by author)

straight ones, do not tend to use such corporeal traces or marks in urban space. For example, there are no signs depicting male bodies or such signifiers in the central city. For most lesbians and gay men this point may seem so self-evident as to be unremarkable: 'it's the closet, stupid!' But I want to bring that thorough-going point into conversation with Lefebvre's remarks. By relying only on visual evidence, we potentially miss an important component to the sexual production of urban space. We also reproduce the standard heterosexuality in Lefebvre's work. So in order to understand sexual–spatial relations better, and to extend Lefebvre's thinking, we must appreciate the significance of the closet.

### *The closet in Christchurch*

Turning to a somewhat different sort of text – the urban landscape itself – we can begin to appreciate the different sexual spaces produced in the city, and consider how the closet became spatialised in them. Historically, the closet has been a necessary structure in the sexual geography of New Zealand. It is important to bear in mind that the very existence of homosexuality was only legalised in New Zealand in 1985 (Atmore 1988; Parkinson 1989; *Broadsheet*, Summer 1993).

*Figure 3.9* Femme Fatale massage parlour. (Photo by author)

More relevant to my discussion here, the space of homosexual acts had been explicitly outlawed until 1986 (*New Zealand Statutes* 1986). Literally, 'Keeping a place of resort for homosexual acts' had also been illegal before that time. The law applied to the managers, tenants and landlords of premises for 'the commission of indecent acts between males' and carried a penalty of up to ten years' imprisonment (*New Zealand Statutes* 1961). The sexual law reform in 1985, however, made homosexuality legal in New Zealand. It attacked the notion of the closet, with an increased visibility of gays and lesbians in New Zealand culture. By 1992 discrimination on the basis of sexual orientation had been outlawed, and the country is one of the few in the world that allows same-sex couples to immigrate.

These sorts of legal sanction help explain why gay venues are so hidden (i.e. closeted) in urban space, compared to their straight counterparts.[16] Despite these laws, gay saunas and bathhouses have long existed in New Zealand. In

Christchurch specifically the first one opened nearly two decades ago. What is so fascinating, however, is the different-yet-similar production of urban space at these venues: [17] 'different' in the sense that one is so self-consciously visible in the landscape while the other is not; 'similar' in the sense that both are instances of productions of urban space working through sexuality. That paradox is explained partially because the closet is a production of space in Christchurch.

Its cultural significance was repeatedly identified and discussed by owners of gay sex-on-site venues. Starting with the visual evidence, typically gay spaces (bars, saunas and cruise clubs) are entirely inconspicuous in the urban landscape. You simply would not know they were there from the streetscape. It is important to note that not all straight venues have such explicit signage. Nevertheless, the point I would stress is that *no* gay venues adopt the explicit visual–sexual strategies that many straight parlours do. Figures 3.10, 3.11 and 3.12, for example, show the entryways to two of the city's gay bars.[18] In each case, the sense of

*Figure 3.10* Alleyway to a closeted gay bar. The former UBQ bar was located down this dark gated alley in a dim stairwell. (Photo by author)

*Figure 3.11* Streetscape around UBQ alleyway. The narrow alley is difficult to find. (Photo by author)

concealment is clearly illustrated. One (Figures 3.10, 3.11) is located at the end of a narrow, gated alleyway on a busy downtown street, easily missed. The other (Figure 3.12) is a nondescript doorway in a row of street-level retail shops: a veritable closet door in the streetscape. The name of the bar, furthermore, suggests a certain anonymity afforded by the closet space's concealment. The same visual concealment obtains for the city's gay bathhouses, as shown in Figures 3.13 and 3.14. Both venues are located on busy downtown streets. Unlike their straight counterparts, however, these premises do not announce their trade in any sexually explicit/visual way. A small sign above a narrow doorway signifies one venue as a health club. The other venue is not identified at all on the exterior. The interior foyer, however, has a sign that simply states 'Private Club'.

Now this sort of queer urban landscape is nothing new for gay bars, to be sure. In their characterisation of Vancouver as an emerging global Pacific Rim city for instance, Ley *et al.* (1995) have explained the increasing visible presence of gay bars by the city's growth and the increased aplomb of its gay community through the 1970s and 1980s. Nevertheless, I want to make the more specific point that this closeted landscape speaks back to Lefebvre in a number of ways. Foremost, it confirms the point that gay space is concealed in the city. To give a sense of just how hidden and inconspicuous these closets are in the city, consider the following anecdote from the owner of one operation:

*Figure 3.12* Streetscape around the former No Names bar. Besides the name, the closet is signified by the inconspicuous doorway amidst the busy urban landscape. (Photo by author)

I don't think many straight people even know we're here. And I can use one example. A taxi driver. . . . And about three-quarters of an hour later he comes back. And I said, 'Oh, I didn't ring a[nother] cab.' He says, 'No, no, no. Is this a gay sauna?' And I said yeah. And he said, 'How long have you been open?' And I said whatever it was at that stage,

*Figure 3.13* Streetscape around gay bathhouse, inner Christchurch. This sex-on-site venue is located amidst several highly visible straight sex-on-site venues. Its closeted presence in the landscape stands in sharp contrast to the highly sexualised images in Figures 3.5 to 3.9. (Photo by author)

*Figure 3.14* Gay bathhouse. (Photo by author)

> 13 years or so. 'Aw like hell!' he says. 'I've been driving in this bloody
> city for 23 years and I never knew you were here!' Now there's an
> example. A taxi driver! And he had never known there was a gay sauna
> in town.

Here, a presumably straight Cantabrian with considerable local knowledge is
surprised to discover a gay sauna in the central city. The closet appears to keep
the space secret and beyond his representational space.

The cultural significance of the closet, however, does not just end at the venue's door. It operates within the venues themselves. One owner made this point as he stressed the importance of secrecy to his clientele:[19]

> Secrecy has always been my top priority. I've known people for 17 years, and I still don't know their names. And I think that's right. I might put them down as 'DG' (dark glasses) or TCH (tall curly hair). Initials like that. I only realised when I was in hospital with cancer three years ago how many people were deeper friends than what I'd thought. Because people come in (head down), pass the money over, and they don't talk to you or acknowledge you're even there. You try not to take much notice of them. But I was getting cards with 'Sum.' on them. That's Sumner [an outer suburb]. And that's what I'd put down in the book for that person. And 'up' uphill Cashmere [another suburb]. And my sister would say, 'Who's that?' And I'd say, oh, that's one of my customers!

Sex successfully takes place in the closet because it is so secret. These men depend on the invisibility and anonymity produced in that space in order to have sex with other men. Another owner went on at length about his decision-making around the signage of the bathhouse. His account demonstrates the complex and secretive coding that is meant to go unseen by straight eyes:

> I guess when we also first started we'd observed from our own travels around the world and in NZ, we'd seen that venues of this nature were very often unmarked. One became – especially as a foreign tourist in a foreign city – one became quite adept at sniffing these places out. There is a certain sameness about the anonymity-yet subtlety of the signage. It would tell a guy looking for a sex-on-site venue that this is – of all of the businesses in the street which one do you think most likely to be it. And they look in our door, they see 'Private Club'. We don't have the name of the business anywhere out there. It just says 'Private Club'. So if somebody looks us up in the phone book and it says 'Lichfield Street', as long as they can find the street and as they walk down the block. I mean, some gay guys come in not even knowing what it was. But they were strolling down the street. They saw our doorway and said to themselves, 'This looks like a family outfit.' And they'd come up. And we've gotten more customers purely because they recognise the signals that say that this is a gay bathhouse or this is a male-only facility.

When asked to discuss these spatial signals, he had a great deal to say about the narrow focus of the codes, designed to attract the attention of men seeking sex with men, while at the same time being completely invisible or innocuous to heteronormative eyes:

> What are the signals? I think 'private club' is a fairly well used term within gay businesses. It has been for years. It's indicated that it's not open to the general public. But at the same time we have the Visa and MasterCard sign in window. So it's saying that it's a private facility but it's also a commercial entity (because of the credit card signs). But we didn't, for instance, put the silhouette of a naked woman there. That would have said 'massage parlour' whether we put the word massage parlour there or not. It's a little difficult to describe it, but I mean to the average straight person walking past, they take a look in. The first thing that might hit their mind is that – if anything hit their mind – might be 'oh, wonder what that is.' By the time they get sort of another couple of paces down the road it's gone. They're thinking of something else. And that's really another reason why we want it that way. We didn't want to have such a hugely elevated profile on site. Whereas we've been able to generate a profile from the media. So a lot of people have heard of us but a hell of a lot don't know where it is. And that's the kind of profile that we want. So the people who actually do want to use our services will take it a bit further, and find out where we are. I guess it's just the sign that gives us that hint of exclusivity that's likely to attract a gay man or a guy who's looking around for a place that's supposed to be private and discreet. That would certainly suggest such a place.

From these comments we can appreciate the very structuration of the closet's production in urban space. It is not simply an external effect imposed upon queers by homophobic straight society (though it is surely that too). Men who have sex with men, as owners and clientele, actively produce the closet themselves through these venues in reaction to heteronormative structures.

The visual evidence offered above, along with these discussions, confirm that gay sex-on-site venues are very much productions of closet space. Their architecture, signage and arrangements are spatial strategies by which urban space is sexualised. Owners and patrons can see beyond the visual (non)-markers of these establishments. Thus in exactly the same part of the city, different sexualities are using rather different visual strategies in their productions of space.

This discussion raises an important point *vis-à-vis* Lefebvre's work. It confirms his suspicion of relying solely on visual epistemologies to identify the production of space despite Lefebvre's heterosexism. Another parlour owner, when discussing the reasons why gay sex-on-site venues remain so closeted despite the central location, used a sort of 'out of sight out of mind' explanation. It confirms the elisions heterosexual structures visually make in producing urban space.

> You know, certain sections just don't know. You can't use an overall effect. Because, like I use the example of when you go to a bookshop. This is why I disagree with the Christians when they say that dirty books shouldn't be on the shelf. If you're looking for a dirty book and you go through three shelves, and you see your dirty book. Ah, that's the one I want. Right. Now, if someone asks you what was any of those other books that your eyes just went past you won't be able to tell them. And it's the same as walking down [a main street]. If you're walking down the street looking for a record shop, and you were asked, you would be able to name maybe one or two shops that you knew, but not all of them. The mind only registers where you're focused basically. I mean, there's the odd person that would know every single shop in the street.

On these grounds Lefebvre might speak back to Sedgwick. If the closet is just 'an open secret', a 'knowing by not knowing', how do we explain the overall ignorance of these venues in the straight geography (in both spaces of representations and representational spaces)? In other words, the subtlety of Sedgwick's definition might not always be appropriate. As a production of space, the closet can also be largely or completely unknown, if knowing is so epistemologically based in vision. The production of gay space in Christchurch, then, underscores the importance of Lefebvre's remarks on vision and the hegemony of ocular ways of knowing. Men who have sex with men ironically rely on the importance of vision in sexualised space to go unseen. This is true in spite of his own heterosexism and essentialising tendencies.

Building on this point, however, it is not just that gay space is closeted, but that the closet space is a part of the very production of inner Christchurch. Here we must bear in mind Lefebvre's remarks on the relations between sexual space and the commodity. This final point may be teased out of the Christchurch case. In discussions with bathhouse and sauna owners, the importance of the closet could not solely be explained by an all-prevailing heterosexism. Recall that, like the straight massage parlours, these venues are commercial spaces. Sex is still a commodified social relation in that space. The clustering of gay bars and saunas in Christchurch – which does not have any 'gay ghetto' – was explained this way through a mixture of local culture and economic rationality:

> Being close to the other gay venues is significant in a place like Christchurch. Because the city of Christchurch is reasonably condensed people tend to park their cars or cab in. And then they are able to walk to all

their entertainment venues. And that's what they expect they'll be able to do. Whereas in Sydney if you wanted to go to a particular place you might end up spending $10 on a cab. Christchurch people are not like that. They're used to spending zero on transport when they go out socialising. Except maybe chip in for a cab into town something like that. So we sort of didn't want to be too far out in the boonies for that reason. We wanted to be near the other gay venues so that people could move from one to the other or go from one to the other with relative ease. Our challenge would be to get them in the door once. But beyond that we introduced to the New Zealand sex-on-site venue scene the concept of being able to come and go as many times as you like. As long as you pay once, you're cool. There's a certain supermarket situation here in terms of volume. And if you can get one guy paying to come in once, the longer period of time he's on site the better. So the clustering is important for us, especially in the niche market that we're operating in. More people there are around who are available to us as potential customers the better. The niche in the purest sense is men who have sex with men. That's our market.

The significant point here, though, is that while both forms of sex-on-site space are always commodified, the production process is rather flexible in Christchurch. Lefebvre's comments seem cogent in explaining the spaces produced by straight venues, but cannot explain the more closeted gay venues. Even though the body is not fragmented and commodified visually (at least in public space), gay venues actively participate in the production of the closet because it is economically rational to do so. Three examples demonstrate this point.

When I first arrived in Christchurch, I looked up the local gay bar in the national gay newspaper, which told me it was the city's most popular gay and lesbian bar. Address in hand, I went looking for it, with no luck. Intrepid geographer that I am, I finally located the bar at the end of a dark, gated alleyway. This was the landscape shown in Figure 3.10, and its closetedness has already been noted, but it goes deeper. At the end of the alley there was a dimly lit stairwell going down to a windowless black door. I pushed the buzzer, heard a garbled voice on the intercom, and then heard the door unlatch. I walked into the basement bar and talked with the owner, who watched me at the door from a one-way camera. He explained the economic advantages to this closeted set-up. It kept homophobes and bashers out, and it made men who were deeply in the closet feel safe. He then asked me if I wished to purchase a swipe-card (renewable annually) that would let me into the pub automatically.

During an interview with a sauna owner, he went on at length to make sure I understood that his ability to produce closet space was the cornerstone of his business success. He discussed this point by contrasting his venue with the straight massage parlour's explicit signage:

> About the difference between us and the straight-sex venues: I mean in our case it was dictated by a lease. It's private. It has nothing to do with the city or external influences. And I can't say that it would have been anything different had that clause[20] not been in the lease. That's one of the clauses that I didn't have any problems at all with. I didn't say, 'Awwww . . . okay.' I wouldn't have gone gaudy or garish anyway. I mean whichever way you go about it there's still an attitude in a lot of people about male-to-male sex. Particularly when it's something like this. We don't really want to incite a public outcry, pushing the limits of social acceptability. I mean, the only reason it's socially acceptable for straight men to do what they do is because prostitution (of the male– female kind) is 'the oldest profession in the world'. So there's almost an acceptability there because of its longevity. Also it's a sexual liaison between two different genders. That's acceptable. But it's the kind of attention that I don't want to go seeking, you know? I don't see it as being in my commercial interests to do that.

The inconspicuous nature of bathhouses in the urban landscape certainly is not unique to Christchurch. Even gay-friendly cities like Vancouver and Seattle have bathhouses and sex clubs with an inconspicuous presence in the urban landscape.[21] Yet such a closeted materiality takes on heightened significance in a homophobic place like Christchurch.

At the risk of speaking for, and (mis)representing the other, the closetedness of lesbian sexuality in Christchurch also illustrates this point.[22] Lesbian space is extremely hidden in Christchurch, much more so than gay space. Traditionally, activists informed me, it has been produced only in the private space of women's homes. Several people mentioned a dinner-party circuit for lesbians in the area, much like the one in 1950s Buffalo (Lapovsky-Kennedy and Davis 1993). To a limited extent lesbians also patronised gay bars in Christchurch – though the intensely gendered space of certain gay bars alienated many women. During my fieldwork, a lesbian bar opened in one of the main streets in downtown Christ-church, sited between two massage parlours. It was marked in the landscape by a series of pink triangles and woman symbols. The venue lasted little more than a year. According to some, its demise was explained because there were not enough lesbians to make the venue profitable; according to others, the inhospit-able patriarchal location made the venue unattractive to any self-respecting feminist-lesbian. For these Cantabrians too, the closet is more than just a metaphor.

How are we to theorise the closet as a part of the production of urban space? The work of geographers Knopp and Harvey allows us to extend Lefebvre's insights, while correcting for his heteronormativity and essentialism. Knopp has been fundamentally interested in the relations between sexuality and capitalism both generally and in terms of urban space specifically. He argues that one cannot understand the spatial organization and representations of sexuality without appreciating their imbrication with class relations specifically, and capitalist relations of production generally. His work on gentrification in New Orleans (Knopp 1990), for example, shows the central role gay men play in reconfiguring the class relations and urban land market in that city. These accumulation processes in the urban land market help dismantle the closet in urban space. Yet he has also been keen to stress there is no inevitable outcome of the interplay between heteronormativity and capitalism. Following Harvey, he has explored more recently the ways in which representations of sexuality have become increasingly significant in the production of urban space (Knopp 1992; 1995a). As he puts it (1992: 661), 'In patriarchal, heterosexist capitalism, investment in the physical and social infrastructures that constitute places is investment in particular configurations of gender and sexual relations as well as in particular production systems and class structures.' Knopp goes on (1995a: 155) to note that it is not simply a unidirectional causality here; it is not simply that gay relations are manifest in urban space only because they are functional to capitalism. Capitalism can also encourage queer consciousness and politics:

> As sexual experiences in particular became increasingly dissected, categorized and commodified (e.g. in the ways Bech describes), the possibility of new (but socially disruptive) sexual experiences being profitably produced also increased. The proliferation of commodified homosexual experience, for example, led to a homosexual *consciousness* among some people, and this was very threatening to heterosexualized gender relations underlying the industrial city [emphasis original].

Here Knopp draws on and extends Harvey's (1985) emphasis on the role of image, especially images of otherness, as important means for capital accumulation. As place-images reflect and reinforce sexualised identities, so too do they sustain capitalist accumulation.

To understand the presence of the closet in Christchurch, I would extend Knopp's point deeper by also drawing in Harvey's (1996) notion of 'spatial fix'. Harvey argues that one of the fundamental contradictions of capitalism is its ironic geography: for capital to accumulate it must be mobile, that is, it must move between places. Yet capital cannot just be mobile; it must also remain fixed in place (for instance in factories, equipment, or longstanding class-relations in a locale). Just as the mobility of capital can fix a crisis of accumulation produced inevitably in a previous round of investment, Harvey reminds us that so too capital's very placement and materialisation – through being in a specific location – can also allow the accumulation process to continue (or even accelerate).

> The produced geographical landscape constituted by fixed and immobile capital is both the crowning glory of past capitalist development and a prison that inhibits the further progress of accumulation precisely because it creates spatial barriers where there were none before. The very production of this landscape, so vital to accumulation, is in the end antithetical to the tearing down of spatial barriers and the annihilation of space by time.
>
> (Harvey 1996: 610)

For Harvey, spatial fixes can occur at any spatial scale: from European empires to regional landscapes, to the very internal structure of cities and neighbourhoods. While Harvey has stressed the crisis-resolution function of spatial fixes and their flexibility in a context of post-Fordism, I simply want to draw out the more modest point that the closet can be thought of as part of the very flexible and variegated arrangements in the production of urban space, which are not simply a function of heteronormativity alone.

If Knopp is right and we must be ever vigilant in recognising the tangled relations of capitalism and sexuality, and Harvey is right to point us towards spatial fixes in the urban landscape, then it becomes theoretically important to see the closet as a particular production of a spatial fix. It is a means by which sexuality can foster capital accumulation. What is more, the closet flexibly allows both productions of space to occur in the *same* location in the city.

The point I would underscore here is that the production of closet space cannot simply be explained by the prevailing structures of heterosexuality *sui generis*. In other words, Lefebvre's insights are helpful in explaining the closet almost in spite of themselves. Spaces of sex are commodified regardless of their orientation. The spatial means and advertising of that commodity, however, are not the same for gay and straight sex-on-site venues. Lefebvre's remarks on the body and fragmentation seem out of place in discussions of the production of gay space, because of the theoretical significance of the closet in structuring sexual relations.

### *Coda*

By the time I left Christchurch in 1998, the landscape of sexuality had both changed and stayed the same, perhaps reflecting the flexibility of its spatial fix.

Two of the more closeted bars had closed, while two new bars opened in the same general area of the inner city. A dance bar (with an open façade on the street) became known as a 'mixed' venue for both straights and queers. A new queer bar also opened, which explicitly tried to cater to both lesbians and gay men. It presented itself with an illuminated rainbow sign, clearly visible at a major city intersection (see Figure. 3.15). Has the closet ceased being a profitable production of urban space, a spatial fix, for gay space in Christchurch? The sex-on-site venues remain hidden as ever, to be sure. The dance bar might be open to the street, but its dim lighting and arrangement of tables make it difficult to see into the bar, but easy to see out from it. The rainbow sign on the newer bar, however, suggests that the closet isn't necessarily a ubiquitous spatial fix for queer urban space. Ironically, there also has been a certain closeting of *both* gay and straight sex-work in Christchurch. There has been an increasing trend towards sex-work to be done not in parlours or on the street, but with a newspaper advertisement and a cell phone. The high overhead costs of working in a massage parlour are often given for the reason for this trend, but so too is the flexibility of that form of labour process.

*Figure 3.15* Bar Particular (upstairs). This new lesbian and gay bar marks its presence in the landscape with a bright, illuminated rainbow-flag sign. Note also the street sign indicating that the intersection is under police-camera surveillance. Could this visibility signal the end of the closeted landscape of inner-city Christchurch? (Photo by author)

## Conclusion

The purpose of this chapter has been to sensitise Lefebvre's work to the existence of alternative sexualities. His thoughts on the role of sexuality in the production of space are insightful, and need to be extended. Taking the central city of Christchurch, we can appreciate the importance of sexuality to the production of space. While his remarks about its production through abstract space seem to be overstated, we can see how capitalism and heterosexuality intertwine to fragment the body in abstract space so that it becomes objectified and commodified. That account, however, can suffer from an extreme heterosexism. In an attempt to widen the parameters of sexual relations, I explored the spatial metaphor of the closet through the Christchurch inner city: a place where sexuality is both ironically and intensively part of the landscape. The discussion here suggests that the closet is more than just a spatial metaphor. It has material expression in the urban landscape. Abstract space and spaces of representations often elide or ignore it – especially compared to straight sex-on-site venues. These sorts of representations, along with representational spaces, can be used to demonstrate the existence of spaces that conceal gay sexuality.

Mapping sex-on-site venues produces a visual representation that shows the production of sexual space throughout the inner city. More narrowly, it shows that gay and straight venues are generally in the same area, but the former is much more concentrated spatially. Visual evidence confirms these patterns and further shows how stealthy gay venues are in the urban landscape. These unseen sexual productions have important implications for Lefebvre's account of the sexual production of space. Principally they suggest that his emphasis on visuality might be somewhat overstated when multiple sexualities are acknowledged. Reading the landscape of the inner city helps explain the processes behind these closeted productions of space. The effects of closet productions are paradoxical, and here Lefebvre raises important geographical issues for queer theory. Unlike straight venues, the closet conceals same-sex desire – but by doing so actually enables its practice. In this way, we might think of it as a kind of spatial fix where sexual relations are commodified. The closet makes male sex-on-site a valuable commodity in the city. Had the sex-on-site venues been more like their heterosexual counterparts in Christchurch, they would not be patronised.

Lefebvre's work on the production of space, then, can be extended to contemporary thinking on the relations between sexuality and space. His writing shows the limits of abstract space to be sure. It exposes the important nexus of sexuality, space and the commodity. And while his work needs a greater appreciation of multiple sexualities, it nevertheless has a significant role to play in introducing space to queer theory. Specifically, it can show how the closet is a material facet of city space.

# Notes

1 For a discussion of this placement more generally see Jacobs 1996.

2 This is perhaps an unfair criticism, given that the Stonewall Revolution occurred a mere five years prior to *The Production*'s publication. Yet by the early 1970s critical theoretical work on homosexuality was challenging the blatant heterosexual epistemologies in French thought. See, for example, Hocquenghem (1993, first published in France in 1972).

3 This point is developed in Chapter 4 more fully with an analysis of the census.

4 The relations between sexuality and capitalism have been consistently explored throughout the history of queer theory. For an early example see Murray 1984. For more recent examples see Morton 1996; Knopp 1992.

5 Jay discusses Lefebvre only briefly in his survey of French thinkers, framing him within a discussion of the Situationist International. See Jay 1993: 418–20.

6 These texts were specifically chosen because they have often been helpful tools in geographers' analysis of urban space.

7 This trend includes the opening of the understated (yet still controversial) Christchurch Casino in 1994, as well as an increase in the number of bars and sidewalk cafes along Oxford Terrace and Cashel Mall. See Canterbury Tourism Council 1995.

8 That abstract spatial productions never stand in cultural isolation is taken up more directly in Chapter 4.

9 This is an especially important point in the context of the AIDS crisis. See Brown (1999) for discussion.

10 The map was produced by using standard spatial measures of central tendency and dispersion. See Clark and Hoskins 1986. Locations were generalised to circles (rather than point-mapping each venue) to protect the concealed nature of the closet. What is more, the standard-distance circles conveniently represent the point that commodified sexual relations exist *generally* in the inner city.

11 The congregation of commercial sex establishments is hardly new, as work on various 'red-light' districts has shown (McNamara 1994). Zoning and cheap rents often combine with factors like accessibility to produce parts of the city given over to public sexuality. In Christchurch, evidence suggests that these factors play a limited role in the clustering of massage parlours, etc., in the inner city. There is no explicit zoning that specifies these venues must locate in certain parts of the city. The law does require, however, that massage parlours conform to local land use codes (*New Zealand Statutes* 1978). Space availability is also a factor, as the central city has overbuilt its office space, making downtown properties particularly attractive. Parlour owners have noted some economic benefit to spatial clustering, but it seems to be more of an accident of history that caused the concentration of parlours in the inner city. It is very difficult to open new parlours, as one might expect. Consequently, certain addresses tend to stay as sex-on-site venues while their names, ownership, or management change. Certain buildings, workers recall, have just 'always been parlours', though they may have had three or four incarnations over the years (Eden 1997). These factors have combined in Christchurch to produce the rather clustered pattern of heterosexual sex-on-site venues in the central city.

12 A referee questioned the accuracy of this claim, given one venue's eroticised mural of two semi-clothed women almost embracing. The aim of the image is to excite

heterosexual men; not to entice or condone lesbian sexuality. See Cooper (1994) on this point.

13  I recognise that gay bars are not primarily venues for sex like bathhouses and saunas. They are, however, spatial expressions of gay sexuality in urban space, and it is on these grounds that I included them in this analysis.

14  Note that this map does not include beats, cottages, or other venues for cruising or gay public sex. Such spaces were rather fluid and shifting at the time of the study. Hagley Park, just to the west of the inner city, has historically been a venue for gay cruising and sex, reaffirming the broader argument in this section of the chapter.

15  I recognise the use of the term 'gay space' is problematic when referring to locations like male bathhouses and saunas, where some patrons may not self-identify as 'gay' themselves (they are simply men who have sex with men, or are bisexual, etc.). I use the term, however, in part because of convenience, in part because gay people also patronise these establishments, and because these are typically coded as gay spaces by locals who are aware of them.

16  To be sure straight massage parlours/bathhouses and gay bathhouses are not exactly alike. There are no sex workers *per se* in the latter. In gay venues, patrons pay a fee to enter the premises; they do not necessarily pay for sex while there. In straight venues, patrons do not necessarily pay a fee for entry into the building (though workers pay 'shift fees'); they pay for specific services. Locally, however, parlours and bathhouses are seen as equivalent and referred to as 'sex-on-site venues'.

17  Of course it is also fascinating that just because there was law reform in New Zealand does not mean that the closet has disappeared. It would be interesting to consider historically the ways in which its form has shifted.

18  These venues have closed since the fieldwork in 1996. Therefore, I have not altered the photos.

19  For purposes of anonymity, I do not cite the names of interviewees specifically. All individuals interviewed work in or around the sex industry in Christchurch, and were interviewed in 1995. For additional aspects of this research, see Brown 1999.

20  The interviewee is referring to a clause in his lease that prevents him from displaying lurid signage on the building.

21  For a discussion of this point generally, see the 'rules for the bathhouse' article at: <http://flexbaths.com/bathhouse101.htm>.

22  The information reported here is from a research project discussed in Brown (1999), as well as more informal local knowledge from queer colleagues and friends. Any errors or omissions are my own, however.

# 4

# NATIONAL CLOSETS

## Governmentality, sexuality and the census

*With Paul Boyle*[1]

### National Coming Out

In Britain, tales of the closet in the nation-state are receiving a great deal of coverage as we research and write this chapter. On 8 November 1998, Agriculture Minister Nick Brown publicly declares his homosexuality after being threatened with exposure by a former lover (Travis 1998). This story comes on the heels of the resignation of Ron Davies, Minister of Welsh Affairs, who reported being mugged on Clapham Common late one evening. The mysterious resignation could only be explained, the media claimed, as pre-empting an outing since a likely reason for his presence on the common at night would be to cruise the 'beat' for gay sex.[2] Still earlier in the month Cabinet member Peter Mandelson was outed on a national news programme. With all these closet doors swinging open, *The Sun* provocatively asks in a front-page headline if the country was being run by 'a gay mafia of politicians, lawyers, palace courtiers and TV bigwigs' (Pallister and Gibbs 1998: 3). And the issue continues to receive press attention, despite the fact that a majority of Britons feel that being gay is morally acceptable.[3]

These news stories raise the possibility and importance of the closet being framed in national space. In each case the closet is situated at the national scale, specifically in the *state*. For each man, the closet allowed a gay presence in the government. Yet it also immediately conjoined the space of the state with the space of the nation, as debate ensued over whether state officials should conceal their homosexuality. Here, of course, we could recognise national government as part of 'the Trinity of the closet' (Signorile 1993; see chapter 1). In what other ways does the closet operate through the nation-state as a material location? To explore this question, I began to discuss it with a visiting colleague, Paul Boyle, who has expertise with the British census. At first glance the combination may seem odd: a gay qualitative cultural geographer and a straight quantitative population geographer appear to have little common ground in the discipline these days (Johnston 1997). Yet we have found the collaboration to be enormously instructive, productive, and at least one result is this chapter. Given our overlapping areas of expertise, we decided to focus on the ways that national censuses might be a closet space. Specifically we examine both the British and US censuses. We take this empirical tack partly because of our training and interests, but also because censuses are typically argued to provide the most reliable picture of the population of a nation-state; the choice of questions in some sense must be a reflection of what's deemed to be important about a nation's inhabitants.

The empirical analysis that follows draws on and speaks to Foucault's (1991) notion of *governmentality*. Briefly stated, the concept signals a decidedly modern form of state power/knowledge. It highlights the interactive power between the knowing (in this case, state bureaucracy) and the known (a nation's inhabitants). Governmentality calls attention to the simple point that the state's administrative apparatuses play a key role in the way nationals come to know themselves as a coherent nation. State power is no longer simply the power to wage war or pass laws, it also lies in very ordinary, mundane bureaucratic practices. Specifically: a state's own knowledge of its population powerfully frames the conditions and terms through which its citizens can see themselves as a nation. In this way, they come to 'govern' themselves through the state's 'mentality'. Such a network of both descending and ascending exercises of power is clearly Foucaultian. And we want to suggest here that national censuses are a clear material example of this practice. More specifically, we want to argue that the census produces the closet in national space through governmentality.

In what follows we chronicle our attempts at seeing the closet in the census. We note in 'Queering the nation' that the census has yet to be considered at the national scale within queer studies. Next, we review the literature on governmentality, arguing that the census is a prime example of such a power/ knowledge practice. This literature, however, has failed to consider sexuality as part of national governmentality. We therefore use the remainder of the chapter to address both literatures in two empirical moves. First in 'Closet governmentality and the census, part 1' we consider macro-level data with the American census, showing how difficult it is to see any queer presence within tract-level data. Turning to individual-level data in part 2, we extend this point by running an experiment using both US and British census micro-data that theoretically should reveal single-sex couples in the nation. In the US the number of same-sex couples is surprisingly low, and these data are laden with assumptions we identify as part of governmentality. In the British context, an interesting story is revealed about the way governmentality emerges through bureaucratic standards of reliability and validity. To our surprise same-sex couples were actually coded out by Britain's ONS (Office of National Statistics), and we explore the underlying reasons for that closeting move. Both examples identify the governmentality of the closet at the national scale. Gays and lesbians in both countries remain difficult, if not impossible, to see through this national image. As Foucault would expect, these instances of national closeting arise not so much from some explicit top-down homophobia. Rather, as is always true of any discourse, the devil is in the details. The 'causes' of this closet are rarefied and multiple: societal heteronormativity, categorical tyrannies, concerns over personal privacy, and statistical rigour.

Saturating this chapter, of course, is the ethical question of whether or not we ought to be able to see lesbians and gays in the closet with the census. At the onset, however, there are two points we would like to underscore. First, we are using publicly available census data in this experiment, which have been released according to strict confidentiality limitations. In other words, these governmentalities already inhabit the public sphere. In this way we want to call attention to the extent of the assumptions one would have to make to see lesbian or gay people through the census. We agree with the paradoxical knowing/not knowing emphasised by Sedgwick (1993) and Fuss (1991), but underscore how this paradoxical knowledge is more than just a metaphor for something else: it is materialised in national geographies sanctioned by the state. And here we call attention to the spatiality of this closet: its location at the nation-state. Second, we'd like to stress that this chapter is not an exercise in 'gay-spotting', outing, or locating gay people. We are interested in whether sexuality and the census conjoin in the mechanisms of governance, and trying to specify how the closet operates through those mechanisms.

## Queering the nation

### Queer studies and the nation

Queer theory has conceptualised the nation-state, and hence the closet inside it, in a variety of ways, but none have considered the role state censuses play in the framing of the nation. Foremost there has been a clear recognition of the state's power to conceal, erase or deny homosexuality through a wide variety of its legal apparatuses (e.g. Adam 1995; Marcus 1992; Mohr 1988; Kinsman 1987). These works identify just how powerful and pervasive the closet is in national legislation. As a top-down exercise of power, 'law' is an efficacious, visible and direct cause of the national closet. For example, several authors have explored the implications of the famous American Supreme Court case *Bowers* v. *Hardwick*, which upheld Georgia's sodomy law, specifically exercised against gay men (Sedgwick 1990; Hunter 1995). The state supreme court invalidated this particular law in 1998, sparking even further discussion of its closeting effects (Sack 1998). Others have explored the Clinton administration's infamous 'don't ask; don't tell' policy *vis-à-vis* gays in the military (Butler 1997b). In the British context, the ramifications of Clause 28 (which forbids local governments to 'promote' homosexuality in any way) and debates over age-of-consent laws have certainly demonstrated the power of the state to closet gays and lesbians nationally (e.g. Wallis 1989; Carter 1995; Tatchell 1991). Still other examples of the national closet are struck in studies of immigration law and practice. Countries like Australia and New Zealand allow same-sex couples to immigrate, for instance. Likewise Binnie's (1997b) recent work on gay international immigration in Europe also highlights the power of the state to forge the closet. The common point here, of course, is that through their rules and regulations, states sanction the presence of gays and lesbians in their national territories. Moreover, they do so in rather complex and often contradictory and inconsistent ways.

The closet is also identified through a reading of literature, film and cultural practices that treats such texts as markers of national culture and consciousness (e.g. Parker *et al.* 1992; Gearing 1997; Edelman 1994). Acts of cultural production in place, then, are recognised as revealing the closet in national space. Debate over the American National Endowment for the Arts (NEA) funding homoerotic material, for example, frames the nation as a cultural space of the closet (Crimp 1994). In Britain similarly, Sinfield (1989) has argued there were tight links between Thatcherism and the closet in national culture by Thatcherism's endorsements of certain kinds of national arts. The issue of state censorship also manifests the closet at the intersection of national culture and state power. Several authors, for instance, have described how Canada Customs routinely bans so-called objectionable material at the border on the basis of its queer focus (Fuller and Blackley 1995; *Forbidden Passages* 1995). Ironically, much of this same erotic material is allowed past national borders when it is bound for 'straight' bookstores, specifying the rather acute heteronormativity of state power. Canada Customs in this case is a powerful closeting force against even NAFTA! On a more activist slant, Queer Nation has invoked the closet as escapable only by direct retaliation against the structurally heteronormative national 'society' (Berland and Freeman 1993). Coding the subject position of queer as a nation itself authorises modes of opposition to homophobia and violence more or less sanctioned by nation-states like Britain and the US. Finally, Duggan (1995), along with a host of other political theorists, argues we must queer the state. As activists, queers must infiltrate and reactivate state power with a queer-positive affirmation. It also means recognising that queers already do inhabit the state, and this can (though not always) be a powerful strategic position to augur social change (Brown 1997; Cooper 1994). As theorists, it means we all must be attuned to the often subtle ways the state is already queered by stressing the progressive possibilities of siting queers within the state apparatus.

From the discussion above, queer studies show us two related themes. First, the nation sees the closet through a wide variety of media and contexts within itself, though certainly a key node is that of the state itself. Second, this encourages us to push on with a search for the variety of ways the closet operates at the national scale. As Bhabba (1990) reminds us, however, there can never be a single authoritative way of knowing the nation. With this point in mind, we want to suggest that census data might be another important way the nation is narrated – narration that manifests the closet at the national scale. Moreover, given Foucault's influence on queer theory's origins (Jagose 1996; Halperin 1995), it is surprising that his work on government has not been drawn on already.

### Governmentality and the census

Foucault (1991: 90) defines governmentality as 'the art of government', while Gordon refers to it as 'governmental rationality'(1991: 1), referring to a way that government thinks about what it does, and the way scholars ought to think about government's power. Both point to a rethinking of state power away from the typical top-down, coercive use of force model. Foucault himself used the term governmentality to signal a shift (c. sixteenth century) in governmental power from sovereign monarchy to a modern administrative, bureaucratic and liberal-democratic state. Here he notices a relative change in praxis about how a state exercises its power. The standard, top-down practice of state power (the coercive use of force and edict) becomes juxtaposed with – if not superseded by – a rather different, more subtle form of power that is more interactive between government and governed. This way of thinking about state power is akin to Giddens' (1987) account of the increasing relevance of surveillance as a form of state authority over its territory. By contrast, however, Foucault is not so much interested in state surveillance as a deliberate, self-conscious and policy-directed reification, rather in the centrelessness and capillaries of the ways normalcy is produced. Nonetheless, both authors agree that politics in modernity works through governmental knowledge and not force alone.

With the rise of modernity, a governmental rationality emerges through policing and disciplining that blurs distinctions between state and society. Here, of course, the very rise of 'society' is noted, as subjects are transformed into democratic citizens and rational economic agents. Likewise, the functions of the state increase and multiply into concerns like public health, medicine and social work. Concomitantly there is a rise of expert knowledges about the nation (e.g. demography, sociology and criminology). Foucault links the broad definition of government (as the practice of the state) with the narrower definition of 'the conduct of conduct: that is to say, a form of activity aiming to shape, guide or affect the conduct of some person or persons' (Gordon 1991: 2). Besides signalling the rising importance of governance over sovereignty, governmentality also signals the operation of state power through its multiple apparatuses and their *saviours*.

Governmentality defines the array and articulation of institutions, procedures, analyses, reflections, calculations and tactics *vis-à-vis* the population. For Foucault, a population has become governmentalised to the extent that it becomes a datum, a field of intervention, an objective of governmental techniques to be known intimately yet still 'at a distance' (Miller and Rose 1990). This distance is achieved through the rise of statistical methods and calculus.[4] Governmentality is a way of knowing the members of a nation both reliably and confidently through the epistemology of science underwriting quantitative measures.

> Whereas statistics had previously worked within the administrative frame and thus in terms of the functioning of sovereignty, it now gradually reveals that population has its own regularities, it own rate of deaths and diseases, its cycles of scarcity, etc.; statistics show that the domain of population involves a range of intrinsic, aggregate effects, phenomena that are irreducible to those of the family, such as epidemics, endemic levels of mortality.
>
> (Hacking 1991: 99)

This knowledge can be used to address social problems by modern state bureaucracies, but it can also lead to a self-disciplining of the population. Foucault referred to this interaction between individual and nation as 'bio-politics', a way of rendering certain bodies 'normal' and others abnormal, deviant, and still others epistemologically (and thus ontologically) impossible. Consider, for instance, the centrality of the 'average' statistic as a way of representing characteristics of the entire population using a Gaussian or normal distribution. As the nation comes to know itself as a coherent entity through the statistics derived from individuals, governmentality not only signals the extent and form of knowledge about national society, but also its role in disciplining orders of normality. This way of thinking about the nation will make, in Gordon's words (1991: 3): 'some form of activity thinkable and practicable both to its practitioners and to those upon whom it was practiced.' Hacking (1991: 181) argues, furthermore, that the rise of statistical epistemologies are crucial to governmentality:

> Statistics has helped determine the form of laws about society and the character of social facts. It has engendered concepts and classifications within the human sciences. Moreover, the collection of statistics has

created, at the least, a great bureaucratic machine. It may think of itself as providing only information, but is itself part of the technology of power in a modern state.

In the national state a *population* is created statistically as the valid form of knowledge comprising highly reliable facts and figures that reveal problems for the state to address but also the means by which they are to be tackled, and the confidence of knowledge to propose solutions. Populations become data that can be statistically described, measured, and hence diagnosed, treated and bettered. Through the complex web of state data on its nation, reliable categorisation provides a way of representing and knowing that channels state power with state knowledge of its citizens (and non-citizens!). Citizens, and their problems, come to be known, defined and recognised along certain descriptive lines, as in Foucault's sense of *savoir*.

Foucault's ideas on governmentality have been deployed in a wide variety of empirical areas of state activity from economic management (O'Malley, Weir and Shearing 1997) to urban and rural planning (Tang 1997; Murdoch 1997a, b), to mapping technologies (Elmer 1996), to public health (Petersen 1997; Osborne 1997), to migration (Abram, Murdoch and Marsden 1998), and even to garbage recycling (Darier 1996). Such a breadth of following is indicative of the subtlety of its power/knowledge. Governmentality is an ordinary, mundane practice of state bureaucracies. As yet, however, it has not been used to explore sexuality *per se*, never mind the closet. That gap is readily filled by this chapter since the census is one of the most obvious and important ways governmentality works and can be identified in the nation-state.

We argue that the census is a key apparatus bearing on the closet, for it is the state's way of seeing its nation 'at a distance'. To substantiate our claim, we explore the dimensions of the closet through an experiment: we try to 'see' sexuality through the British and American censuses. Neither nation directly represents gay people through its census, demonstrating how the closet operates at the national scale specifically through governmentality. This governmentality, however, has a subtlety that we also demonstrate. Strictly speaking, only certain lesbian and gay people can be seen in the censuses, but only if gross assumptions are made about them. Moreover, there are particular logistic and practical difficulties in the British census that we discuss. The overall point is to show the simultaneous simplicity and complexities of the national closet. On the one hand it is a simple issue: there is no question on the census about sexual orientation. Given the thorough heteronormativity in society, this is hardly surprising. On the other, it is an example of a population's self-disciplining: citizens fill out the census form and researchers and government agents interpret the data. After all, gays and lesbians obviously are part of the British and American nations.

According to the United Nations, national censuses are 'the total process of collecting, compiling and publishing demographic, economic, and social data pertaining at a specified time or times, to all persons in a country or delimited territory' (quoted in Brown 1976). While sexual orientation is clearly not a requirement according to this definition, neither is it precluded. At a practical level censuses serve a number of functions. Most obviously, they act as the benchmark for identifying and understanding changes in national-level or small-area populations. They are used as the baseline for population estimates and projections, and local government revenue-transfers as well as for representational equity in legislatures. They are also used for identifying areas of need, clusters of deprived people, targeting population sub-groups (e.g. ethnic groups in health campaigns, etc.). In a very practical sense, the census is undoubtedly part of the art of government in the modern era.

Developing this point, we would highlight the fact that a basic source of a census' governmental power lies in the power of its scientific epistemology. Its power-as-knowledge derives from at least two scientific principles it embodies. Its reliability is guaranteed by its periodicity (that it happens at regular intervals), while its validity is achieved through its universality (comprehensive coverage) because, unlike other national surveys, it is compulsory (you can be fined or imprisoned for ignoring the census in both countries – and of course the state has the legal means to enforce compliance).[5] The census therefore has enormous epistemological purchase because the state can know the population with such authority. What is more, it is not simply a top-down exercise of power. Since the census relies on self-reported information, nationals take an active, albeit already disciplined, role in their governance. The census, then, can be understood as a tool in the *savoir* of governmentality, a statistical linchpin between the state's power and confident knowledge of the nation. Moreover, because the data are more or less publicly available, census data isn't simply the state's knowledge: it is everyone's. People see themselves through the census. This point is often only acute when state *savoir* does not harmonise with individual nationals' self-identities. Take, for example, the category 'Hispanic' in the American census. It is designed to identify ethnic affiliation based primarily on the commonality of speaking Spanish. Yet it corrals such ethnically and culturally diverse identities as Mexicans, Caribbeans, Latin and South Americans (Gonzales 1997).

Few scholars, however, have gone further to identify specific government statistical techniques of data collection and dissemination as clear examples of governmentality (Miller and Rose 1990). Perhaps the best example is Owen's (1996) historical analysis of the Egyptian census of the early twentieth century to illustrate how its changing categories of families and individuals reflected and reinforced shifts in the agrarian economy. Another involves looking at the use of intra-national migration statistics in local planning and decision-making. Abram *et al.* (1998: 238) underscore the importance of government statistics to the exercise of governmentality when they argue:

> The collection of statistics and the proliferation of inscriptions, with their technologies for classifying and enumerating, thus become effective techniques of governmentality, allowing civil domains to be rendered visible, calculable and, therefore, governable. Furthermore, they herald the advent of subjects who remain 'free' but come to calculate themselves in terms derived from the tools and techniques of governmentality.

So as an instance of governmentality the census provides a means of citizens' self-definition, categorisation that derives from and feeds into state authority. Since individuals in the population actively participate in the census process by completing the form, the census is not simply a top-down exercise of state authority. Questions and categories are offered, but so too are they (re)interpreted by respondents. They are ways of seeing a nation's population done by and for the state to solve practical problems. Where better than the census to look for what the American or British populations 'are like', and where better to acknowledge Foucault's point? The dimensions of likeness, of course, are contingent upon the questions that the state decides to include on the form, the range of values that any such variable could take, as well as the way in which the enumerated interpret and understand what is asked of them.

## Closet governmentality and the census, part 1

So how does the census produce the closet at the national scale through governmentality? As we noted above, the simple question 'Can we see sexuality in the census?' immediately prompts an equally simple reply: no. Neither British nor American census forms include an explicit question on sexuality. Table 4.1 illustrates this point well. It compares the categories derived from questions on sex, marriage, family type and household composition in the US and Britain, questions where one's sexuality would most likely be at issue. Consequently, the issue of sexuality in any governmental knowledge or planning derived throughout the census would be non-decided. It could not be a significant variable, because it is not a variable in the first place. So there is a fundamental presumption of heterosexuality in both censuses that would be reproduced through any use of the data, and heterosexist presumption, of course, is a clear example of the closet at work. This is surprising if we think that gay individuals (and their living choices) are different from the general population, choosing different types of accommodation in different places for different reasons (e.g. Bouthillette 1997). With information

*Table 4.1* Comparison of selected 1990 US and 1991 British census questions and possible responses

| United States census | British census |
|---|---|
| *1 Sex*<br>Male<br>Female | *1 Sex*<br>Male<br>Female |
| *2 Marital status*<br>Married<br>Non-married<br>Widowed<br>Divorced<br>Separated<br>Never married | *2 Marital status*<br>Single (never married)<br>Married (first marriage)<br>Remarried<br>Divorced (decree absolute)<br>Widowed |
| *3 Respondent's relationship to householder (related)*<br>Husband/wife<br>Natural born/adopted son or daughter<br>Brother/sister<br>Father/mother<br>Grandchild<br>Other relative | *3 Respondent's relationship to head of household*<br>Husband/wife<br>Living together<br>Son/daughter<br>Other relative (specify):<br>Unrelated |
| *4 Respondent's relationship to householder (unrelated)*<br>Roomer/boarder/foster child<br>Housemate/roommate<br>Married partner<br>Other non-relative | |

*Sources:* 1990 US census, 1991 British census.

on sexual orientation, wouldn't planners be able to plan better? Yet in other ways it is all too unsurprising. Gay people are thought to be a small percentage of the overall population, and therefore do not need to be specifically addressed in the census. As well, several social characteristics besides sexuality are not asked on the census. In other words, sexuality is part of a constellation of identities that are unknowable through census governmentality. For example, in 1991 the British census does not ask questions on religion, ancestry or income, which are included on the US census. Conversely, details about housing tenure seem to be a rather difficult identity to isolate from American census data, compared to the British census. Important as these variables are, there are two major considerations that are debated decennially within the context of the modern bureaucratic state: the financial cost of adding new questions and the worry that these questions would be unreliably answered or affect the answering of other questions. Despite the potential usefulness of income information in decisions[6] and a vocal lobby in favour of adding it to the British 2001 census, there is still debate about whether it should be added. Clearly then, if the census is conceptualised – and evidence suggests that it surely is – as a mirror by which the nation represents itself, lesbians and gays go unseen. They remain hidden, concealed, in the national closet. They do not belong to the nation, even though we know (through other epistemologies) that they are here. This point affirms the paradox that is signalled by Sedgwick and Fuss. The closet is a knowing by not knowing. But here we would suggest the reverse: for the census it is a not-knowing by knowing too.

Rather than making this the end of the question, however, we choose to see it as the start. For if Foucault's arguments energise our thinking about the closet, then we are spurred to draw a distinction between accuracy and precision. It is accurate to say that sexuality is stealth in the census, but is that precise? From this sort of starting point, we can begin to exercise the subtle ways the closet operates. Notice, for instance, the variable 'Respondent's Relationship to Householder (Unrelated)' (item no. 4) in the American census in Table 4.1. Presumably the value 'unmarried partner' or 'other non-relative' could capture gay men or lesbians who were in relationships with householders of the same sex. Indeed, the American census allows for such a coding. From its own handbook:

> An unmarried-partner household is a household other than a 'married-couple household' that includes a house-holder and an 'unmarried part-ner'. *An 'unmarried partner' can be of the same sex or of the opposite sex of the householder.* An 'unmarried partner' in an 'unmarried partner household' is an adult who is unrelated to the householder, but shares living quarters and has a close personal relationship with the householder.
>
> (US census 1990: b-15; emphasis added)

Here, it would seem, the closet breaks down, at least for the American census – and at least for certain lesbians and gays (see below). The data can be reported and coded in such a way as to reveal same-sex couples. When we turn to the actual data, however, the closet seems to manifest itself again. To test their revelatory power, demographic data were collected on census tracts from a well-recognised gay neighbourhood in Seattle Washington, Capitol Hill.[7] These data were accessed from the census' homepage, which is the most readily available source of census data for the public (http://venus.census.gov/cdrom/lookup/). It is also free data, available at the public library, for instance. These attributes are an important part of the exercise, since the whole idea of governmentality is that power lies in the capacity to represent the population to itself. If governmentality

98

is predicated on a bio-politic of self-discipline, part of those politics also lay in the fact that the population can, in turn, see itself through these data.

Tables 4.2 to 4.6 give a brief statistical summary of Capitol Hill, Seattle, in comparison to the rest of the city and the nation overall, for individuals and households respectively. We can see that it has rather odd or extreme values on several variables that mark it as a decidedly 'different' place from the city of Seattle and the United States overall. The area is more male (54 per cent of residents) than the city or country percentages (roughly 49 per cent), as illustrated by Table 4.2. Such a fact might be relevant given that gays and lesbians desire people of the same sex, so single-sex concentrations in space might suggest a gay neighbourhood. In addition, the ways individuals arrange themselves in household units may provide some clues. Table 4.3 demonstrates that the majority of people living on Capitol Hill live in 'non-family households' (almost 80 per cent, in fact). Specifically, around a quarter of both men and women on the Hill live by themselves (25 per cent and 23 per cent respectively). This statistic stands in marked contrast with Seattle overall and the United States. In those other places, the majority of people live in traditional, family households (65 per cent and 84 per cent respectively). On Capitol Hill, only 23 per cent of residents live in such a conventional arrangement. Since lesbians and gays cannot marry in Washington State, we might also use data on marital status to infer a queer presence in the neighbourhood. Table 4.4 tells us that Capitol Hill is a space of single people. Thirty-seven per cent of men and 25 per cent of women (aged 15 or older) in the neighbourhood have never been married. These are rather higher than the statistics for the city or the nation. Fewer than 10 per cent of either women or men on Capitol Hill are married and living with their spouses there. If we move away from data on individuals in the neighbourhood, and look at its households, similar stories are told. The percentage of households that the census

*Table 4.2* Census summary data by persons: Capitol Hill, Seattle, USA

|  | Capitol Hill | Seattle, WA | United States |
|---|---|---|---|
| *Totals* |  |  |  |
| Persons | 13,117 | 516,259 | 248,709,873 |
| Families | 1,254 | 113,856 | 65,049,428 |
| Households | 9,162 | 236,908 | 91,993,582 |
|  | *% Persons* |  |  |
| *Sex* |  |  |  |
| Males | 54.23 | 48.75 | 48.72 |
| Females | 45.77 | 51.25 | 51.28 |

*Source:* US census, 1990.

*Table 4.3* Percentage of people in each area by household type and relationship

| Household type | Capitol Hill | Seattle, WA | United States |
|---|---|---|---|
| *In family household* | 22.99 | 64.73 | 84.1 |
| Householder | 9.56 | 22.05 | 26.15 |
| Spouse | 6.93 | 16.88 | 20.73 |
| Natural born or adopted child | 3.80 | 19.21 | 29.41 |
| Stepchild | 0.06 | 0.73 | 1.57 |
| Grandchild | 0.25 | 1.12 | 1.67 |
| Other relatives | 1.46 | 2.77 | 3.01 |
| Non-relatives | 0.93 | 1.97 | 1.56 |
| *In non-family household* | 77.02 | 35.27 | 15.9 |
| Male householder, living alone | 24.88 | 7.71 | 3.55 |
| Male householder, not living alone | 7.33 | 3.10 | 1.09 |
| Female householder, living alone | 22.57 | 10.52 | 5.47 |
| Female householder, not living alone | 5.52 | 2.50 | 0.73 |
| Non-relatives | 15.16 | 7.37 | 2.38 |
| In group quarters | 1.56 | 4.07 | 2.68 |

*Source:* US census, 1990.

*Table 4.4* Sex by marital status: percentage of persons in each area 15 years or older

| | Capitol Hill | Seattle, WA | USA |
|---|---|---|---|
| Men never married | 37.06 | 20.19 | 14.45 |
| Married men, spouse present | 7.10 | 19.94 | 28.86 |
| Separated men | 0.96 | 0.93 | 0.93 |
| Married men, spouse not present | 1.45 | 1.13 | 1.14 |
| Widowed men | 0.69 | 1.16 | 1.18 |
| Divorced men | 6.85 | 5.12 | 3.47 |
| Women never married | 25.36 | 16.61 | 12.01 |
| Married women, spouse present | 7.11 | 19.85 | 26.68 |
| Separated women | 1.55 | 1.06 | 1.34 |
| Married women, spouse not present | 0.52 | 0.75 | 0.86 |
| Widowed women | 3.41 | 6.21 | 6.21 |
| Divorced women | 7.95 | 7.05 | 4.87 |

*Source:* US census, 1990.

would define as 'family household' (i.e. the presence of marriage arrangement with or without children) is 14 per cent for Capitol Hill, according to Table 4.5, much lower than the city or the country overall. Nearly half of the households in Seattle are 'family households' (48 per cent), and well over half – nearly three-quarters – of American households are described that way (71 per cent).

*Table 4.5* Census summary data by households: Capitol Hill, Seattle, USA

|  | *Capitol Hill* | *Seattle, WA* | *United States* |
|---|---|---|---|
| Households | 9,162 | 236,908 | 91,993,582 |
|  | *% Households in area* | | |
| *Family households* | 13.69 | 48.06 | 70.72 |
| Married with children | 1.29 | 13.34 | 26.33 |
| Married without children | 8.30 | 23.27 | 29.89 |
| Male householder, no wife present | 1.58 | 2.82 | 3.21 |
| Female householder, no husband present | 2.52 | 8.63 | 11.29 |
| *Non-family households* | 86.31 | 51.94 | 29.29 |

*Source:* US census, 1990.

*Table 4.6* Persons in household as a percentage of households in area

|  | *Capitol Hill* | *Seattle, WA* | *USA* |
|---|---|---|---|
| Single person household | 67.92 | 39.74 | 24.37 |
| 2 person household | 26.16 | 33.98 | 31.94 |
| 3 person household | 4.19 | 12.58 | 17.35 |
| 4 person household | 1.14 | 8.14 | 15.17 |
| 5 person household | 0.11 | 3.37 | 7.01 |
| 6 person household | 0.36 | 1.31 | 2.52 |
| 7 or more person household | 0.12 | 0.88 | 1.64 |

*Source:* US census, 1990.

Conversely, most of the households on Capitol Hill are one-person households (Table 4.6).

So does this census profile frame an epistemology of the closet for that piece of the nation? Facts, of course, do not simply speak for themselves. And this is where we begin to see the complexity of the governmentality in the census. It must work through other epistemologies that we have juxtaposed with these government data. Indeed the very original set-up of this experiment is based on our own experiential epistemology: 'we know it's gay because Michael's gay and he's been there'.[8] The facts only reveal gays and lesbians if we also draw on other epistemologies through which we already know them. To run through the tables again – and here we must suspend political correctness and follow stereotypes, preconceived notions, assumptions, etc. (the 'other epistemologies' washing across the facts to make them sensible) – Table 4.2 suggests Capitol Hill is a gay neighbourhood because it tends to be concentrated male space. Capitol Hill is a queer area according to Tables 4.3 to 4.6 because we know that gays and lesbians cannot marry same-sex partners, so they do not conform to census definitions of family households; they are legally 'single'. The potential limits of the truth we are trying to tell here, of course, can easily be demarcated because other identities can also be suggested by these facts. In other words, different and perhaps equally plausible conclusions could be drawn from these data. For example, Capitol Hill might be interpreted as a place of young single people who might all be straight. Yet recall that our point is not to use the census to claim Capitol Hill definitely as a gay space. Rather it is to see how the closet operates through the census.

On that tack, we would underscore the immediate conjoining of alternative epistemologies with the census because our reliance on them to confirm 'what we already know' shows an interesting complexity to the closet. To put it succinctly, if we bring certain gays and lesbians out of the closet with the census, we do it by simultaneously closeting others with our biases and stereotypes. To know Capitol Hill as a queer space, we must know it as a gay-male space and closet lesbians. To see gays and lesbians because there is a majority of non-family households in the area conversely closets those gays and lesbians who do live in family households. Not all gays and lesbians are single, not all have never been married. Some are children, stepchildren or grandchildren and so on. If we infer that Capitol Hill is a gay area because most households are populated by individuals, we likewise closet those gays and lesbians who are partnered but do not live with their partners in the same household. The point we are trying to make here is that even when we use the census against the governmentality of the closet (in it already because there is no question to capture same-sex situations), we wind up re-instanciating it. We cannot break down the closet without inevitably reconstructing it for somebody else. To our minds this point affirms the 'knowing by not knowing' argument put forth by Sedgwick, but in a thoroughly spatial and complex way. In the end, only certain types of gay people are seen through these census data, the rest remain closeted. Moreover, the validity of this epistemology is quite precarious. This is a form of closet governmentality: a knowing by not knowing – or perhaps more accurately, a not-knowing by knowing.

The complexity of the closet in the census can *also* be understood through the power of scientific and bureaucratic rationalities that reflect and reinforce the power of the census to narrate the nation truthfully. Recall from page 98 that the proper way for an American in a same-sex-cohabiting couple to describe their relationship using the categories in Table 4.1 is as 'an unrelated household member' who is an 'unmarried partner'. Consider how the categories offered in Table 4.1 were translated into the census output chronicled in Tables 4.2 to 4.6. You can see that there is no such category as 'unmarried partner' in the output. The one category gays and lesbians should use to describe themselves when filling out the form disappears when they go to the publicly available data to see themselves.[9] To our minds, this is another complex manifestation of the closet nationally. We would explain it, moreover, not through some intentionally homophobic machination, but rather through the force of governmentality operating through discourses that sustain the truth and power of the census. This closet could be due to the tendency of the variable to reveal small numbers, which would threaten the anonymity and confidentiality guaranteed by the census. Cell counts are suppressed or aggregated in published data when there is a possibility of identifying an individual. Here, the closet appears as a consequence of the census' attempts to live up to important principles that help maintain its own authority and power. If Americans could be identified individually through the census, they would be far less likely to participate willingly in its surveillance, or tell it their truths. Another possible reason might be that the Census Bureau simply did not think the data on unmarried partners (some of whom would be same-sex) or the specifics of non-family household composition would be of interest to the general public. The presentation of data in science is always a process of selection and editing. If the census did offer everything it could, no doubt its public accessibility would be compromised by the sheer avalanche of possible information. Either way an upshot of this erasure is that governmentality's positive emphasis on normalcy is noticeable. The categories 'roomer/border/foster child', 'housemate/roommate', 'other non-relative' and 'unmarried partner' have all been collapsed into the category 'not living alone'. Overall, we would argue that a complex closet is operating through the collapse of categories, done not so much to erase identities but to protect their privacy and make the data accessible.

A large part of the problem, of course, is that the data in the tables above are provided at the census tract, rather than the individual, level. Hence any attempt at reading individuals from that spatial scale belies the ecological fallacy (Babbie 1997). This is the problem of generalising from aggregate-level data (in this case census-tract scale) to individuals inhabiting those numbers. For instance, we can know or calculate the number of men who have never been married, and we also know the number of men who live in non-family households, but we cannot logically infer that they are necessarily the same individuals. It is a recurrent problem for those dealing with readily available census data, because these data are provided for areal units in order to protect the anonymity and confidentiality of respondents (a concern which sustains its governmental power through its power to tell the truth). For that reason, we turned to individual, micro-level data, which are taken from a sample of the population, and whose geographic identification is not as precise as the census tract.[10] The benefit of these data is that we can cross-tabulate characteristics of individuals with ease, thus escaping the ecological fallacy. The cost, as it were, is that these data are not necessarily widely available or readily manipulated without a certain degree of familiarity with the census. Nonetheless, since bureaucratic experts (like academics!) are part of the exercise of governmentality, so the validity of the exercise remains largely intact.

## Closet governmentality and the census, part 2

The origins of this second part of this experiment came from Paul's ongoing cross-national research project investigating whether or not women's labour-market status suffers as a result of family migration (in Britain and the US), which usually stems from a change in male partner's job circumstances.[11] A requirement of the analysis in this project was the identification of 'linked partners' in the respective micro-data samples: the 1991 British Sample of Anonymised Records (SAR) and the 1990 US Public Use Micro-data Samples (PUMS). Both the SAR and the PUMS are individual-level data sets taken from a small sample of the entire national population recorded in the census. Each individual's exact geographic location is withheld, however, so confidentiality and anonymity are retained while the ecological fallacy can be avoided.[12] In both data sets linked partners can only be identified reliably in couples that include a self-declared householder.[13] This is because in both censuses household relationships are structured around the head of household, with individuals identifying their relationship to this person on the census form. The resulting couples extracted in the 'tied migration' project were therefore either married or cohabiting couples involving a head of household. Other married or cohabiting couples not involving the householder in multi-couple households could not be reliably identified. In theory, of course, we would expect to be able to identify same-sex couples in both these data sets, so long as gays and lesbians had self-identified themselves as cohabiting with someone of the same sex. So here, it seemed, even though there was no explicit question on sexuality, gays and lesbians who were in household relationships might be a visible part of the British and American nations, according to the census. And it was at this stage that Paul and I began discussing the possibility that the census might not be so much of a closet after all.

### The US

The 1990 American census has more questions than its British equivalent. As a result the PUMS are more comprehensive than the SAR in terms of variable content and are presented in a much less refined format than the SAR data. The benefit of this 'rawness' is that the PUMS data can be manipulated to create variables that match the information from the SAR (Boyle *et al.* 1999). Although the relationships within households are defined slightly differently in the US census, it is still possible hypothetically to identify same-sex couples. In much the same way as the British census, individuals must state their relationship to the householder.

We utilised a 0.1 per cent random sample of the total population drawn from the 5 per cent sample of the PUMS data. From this, 59,978 individuals or 29,989 couples were extracted, one of whom was the householder. Selection criteria included nuclear family households, partnered adults, not in armed forces, not living in institutions, permanent residents, and persons aged 16–60. We drew information on marital status and sex first. Table 4.7 cross-tabulates marital status (for opposite-sex couples) and sex (for same-sex couples) against region, demonstrating that while the majority of individuals were in conventional, heterosexual relationships, 74 men and 64 women self-identified themselves in same-sex partnerships. So, crudely estimated, there would have been 138,000 individuals identifying themselves in same-sex couples conforming to the selection criteria established above in the US, according to the 1990 census. Even if we make the terribly problematic assumption that gay individuals may be less likely to live in partnerships than heterosexual individuals, the extremely low percentage of individuals living together in same-sex couples as a percentage of total persons in the US (0.055 per cent) seems remarkably small and contradicts cultural assumptions about the very presence of gays and lesbians in the country. The oft-quoted statistic (from Kinsey) is that 10 per cent of the population is homosexual. Even if we reject this benchmark for the lower 2 per cent estimate that has been suggested elsewhere (e.g. Laumann et al. 1994), that would put the 1990 lesbian and gay US population at 4,974,197 which in turn (according to the PUMS data) would mean that only 2.77 per cent of gays and lesbians are living in same-sex relationships.[14]

Table 4.7  Couples in households by region (%)

|  | Individuals in opposite-sex couples | | Individuals in same-sex couples | |
| --- | --- | --- | --- | --- |
|  | Married | Cohabiting | Men | Women |
|  | n = 56,712 | n = 3,128 | n = 74 | n = 64 |
| New England | 3,126 (93.5) | 204 (6.1) | 8 (0.2) | 6 (0.2) |
| Middle Atlantic | 8,826 (94.1) | 524 (5.6) | 14 (0.1) | 12 (0.1) |
| East North Central | 10,428 (94.8) | 556 (5.1) | 4 (0.0) | 10 (0.1) |
| West North Central | 4,690 (95.9) | 200 (4.1) |  | 2 (0.0) |
| South Atlantic | 9,608 (94.9) | 498 (4.9) | 12 (0.1) | 10 (0.1) |
| East South Central | 3,632 (95.5) | 170 (4.5) |  | 2 (0.1) |
| West South Central | 6,316 (95.6) | 274 (4.1) | 10 (0.2) | 6 (0.1) |
| Mountain | 3,104 (93.8) | 198 (6.0) | 2 (0.1) | 4 (0.1) |
| Pacific | 6,982 (92.8) | 504 (6.7) | 24 (0.3) | 12 (0.2) |

Source: 1990 PUMS.

Again, however, we immediately confront the problem of assumptions and alternative epistemologies at work in reading these data, even if they did chime with our less scientific epistemology. To our minds they are also part of the power/knowledge effects of the closet in a national context of governmentality. Consider the assumptions and caveats that had to be made in the above exercise. First, we are relying on a self-reported survey by using census data. Fears over homophobia might very well have cautioned certain gay couples from describing their relationship to the government and ultimately the rest of the nation. This certainly plays into the governmentality of the closet, since people would essentially be closeting themselves on the census form. For argument's sake, what if a question on sexuality was included on the census form? The closet would still be at work since many lesbians and gays would no doubt be highly reluctant to reveal their 'private' sexual orientations to a state authority. That concealment would, of course, be especially apposite to those who would self-identify as already being in some form of the closet (for instance, married men who are bisexual and have sex with other men)! Second, of course, couples who simply appear as two women or two men living together in the same household might be closeted by being subsumed into a population of roommates.[15] Third, we are picking up couples who reside within a single household. With both of these assumptions, a powerful conventional heterosexism is at work. Here gay couples are presumed to ape the arrangements of their straight counterparts. Yet a decided thrust of queer theory has been to question and restructure the hegemony of those arrangements in the first place (Weston 1991; Benkov 1994; cf. Bawer 1996). Do relationships need to be hierarchical? If they do, can the hierarchy shift between partners? To what extent do assumptions about monogamy infect the very category of 'couple' and its heterosexual bias? Do members of a couple need to live in the same household, at the same address?[16] The crux of the census is a record of the population and how it is distributed across housing units. In other words, the data structure affects the story the data can inevitably tell. Fourth, and perhaps most obviously, we are only detecting gay and lesbian *couples*. If there were that hypothetical question on sexuality, this point would not be a problem. However, we have no way of revealing a whole series of gay men and lesbians who help comprise the nation. What about those 97.23 per cent of gays and lesbians – if we believe the controversial 2 per cent estimate – who are single? We simply have no way of detecting them through these data. They remain closeted relative to those revealed above in the sense that the data that show them also work to conceal others. Likewise, what about those queers who might have multiple partners or alternative forms of family? Bisexuals also suffer the effects of the closet since as a sort of categorical *pharmakon*, they are both categories but neither. The point here is that each set of assumptions above creates a closet in order to reveal a small percentage of the population as gay or lesbian. It is fascinating that in order for them to be revealed, we must closet others. This paradox is at the core of governmentality: to see some lesbians and gay men in the nation we must inevitably closet others because of our partial assumptions. Additionally, it is interesting to note that the deconstruction above reveals how the census might be concealing certain numbers of opposite-sex couples too!

If we're interested in the geography of the closet subnationally, we can identify the regional location of these couples.[17] Table 4.7 problematically locates these individuals regionally.[18] Again, truths seem hard to reconcile with our alternative epistemologies of queers in the US. Moreover, statistically there is a problem in inferring any grand truths from these figures. Does it make sense, and whatever could it mean, for instance, that there 'are' twice as many partnered lesbians in the Middle Atlantic region of the US than in New England? The very small numbers make it hard to say anything conclusive even at a regional scale. In order to assume these data are the truth, what other truths do we have to suspend? How much closer are we to revealing the lesbian or gay presence in the American nation?

## Britain

As in the 1990 United States PUMS, it is possible to extract linked partners from the 1991 British SAR.[19] Selection criteria were the same as in the US sample and a relatively large sample of 164,496 linked partners was drawn (Table 4.8). However, to our initial surprise, none of the couples were the same sex (Table 4.9), provoking the question why? Of course, the power of the closet itself

Table 4.8 Marital status by British regions (percentages in brackets)

|  | Married | Cohabiting |
| --- | --- | --- |
|  | n = 145,174 | n = 19,322 |
| North | 8,104 (89.7) | 930 (10.3) |
| Yorkshire and Humberside | 13,208 (88.2) | 1,774 (11.8) |
| East Midlands | 11,406 (87.7) | 1,632 (12.5) |
| East Anglia | 5,704 (87.7) | 802 (12.3) |
| Inner London | 3,706 (79.1) | 982 (20.9) |
| Outer London | 10,436 (86.2) | 1,664 (13.8) |
| Rest of South East | 30,024 (87.4) | 4,312 (12.6) |
| South West | 12,118 (88.0) | 1,646 (12.0) |
| West Midlands | 13,992 (89.1) | 1,714 (10.9) |
| North West | 16,032 (88.9) | 2,000 (11.1) |
| Wales | 7,138 (90.4) | 756 (9.6) |
| Scotland | 13,306 (92.3) | 1,108 (7.7) |

Source: 1991 SAR.

*Table 4.9* Head of household and partner by sex

| | Head of household | |
| --- | --- | --- |
| | *Male* | *Female* |
| | n = 74,531 | n = 7,717 |
| Male partner | n.a. | 7,717 |
| Female partner | 74,531 | n.a. |

*Source:* 1991 SAR.

might discourage couples from identifying themselves as gay, despite repeated assurances that census information is confidential.[20] Presumably this would be more important for those living completely in the closet than for those who are 'out', living together as couples. Certainly we would expect some gay partners to acknowledge their relationship with a householder of the same sex. Thus, while we might expect the number of gay couples identified in the census to be smaller than the true figure, we would not expect it to be zero! Our initial assumption, that we had made some kind of error extracting the data, was proved to be unfounded and this prompted us to investigate this anomaly further. Unbeknown to us the issue of same-sex couples had initiated a debate prior to the 1991 census as a decision had been taken at an early stage of the preliminary census planning process to *recode* individuals who recorded themselves as cohabiting with, or married to, a householder of the same sex. The exact process is described below.

The results immediately raise the question of whether this was some sort of plot by the Office of Population Censuses and Surveys (OPCS),[21] or the government department under which it served, to 'hide' gay individuals. In fact, discussions with census statisticians seemed to reveal a less guileful and more practical reason had influenced the committee meeting discussions that resulted in this decision, which resonate with the subtlety of governmentality. The problem was twofold, according to the census offices. First, the relationship question on the census form provided no guidelines about how gay or lesbian individuals should complete the form. That absence could create a problem of interpretation when lesbians and gay people (at least those in couples) filled out the form. The fear was that while some gay couples would view themselves as 'cohabiting' others might not regard this term as relevant to their status, and without additional information clarifying this point, the resulting information would therefore be unreliable. According to the ONS the decision was based less on the sensitive issue of sexuality, and more on the concern that the information that would result from the returns would be incorrect. The second problem was also of a more practical nature. The numerous computer programs that had already been written to handle the census information, according to ONS, would need to have been edited and there was a series of programs that would have needed changing. It was not possible simply to alter a single line or chunk of code to allow same-sex couples who self-identified as married or cohabiting to remain. This intricate reprogramming would be expensive and risky, it was argued. Also, the decision to prevent gay couples from being identified in the census output had been agreed by the time that the draft census form was submitted to Parliament for ratification. The story does not end there, however. A key figure in the census offices only became aware of, or realised the implications of, this decision subsequent to this time. He reopened the debate and, while he was unconvinced that the decision was correct, he was persuaded by the argument that the identification of gay couples should not be allowed if the wording on the census form could not be changed. This was impossible in the time frame of the census machine, as altering the wording of any part of the census form required that it be resubmitted to Parliament, causing impossible delays and prohibitive costs.

The decision stood, although some attempt was made to investigate the potential reliability of such information if it had been allowed to remain in the census output. A great deal of effort is expended prior to the census proper to make sure that the data are collected reliably. The 1989 census test included the same relationship question as used in the 1991 census (Table 4.1) and a sample of these forms was examined to quantify the number of same-sex couples. Of the 7,500 households that were checked, only one 'genuine' case occurred. There were 36 additional cases where the couples were recorded as being of the same sex, but when examined further these all turned out to be errors such as people completing the form incorrectly or incorrect sex imputation by the census offices (see below). Prior to the census, then, it appeared that the information on same-sex couples that could be derived from this question would probably be unreliable.

The processing of the actual census returns is a computerised process, but clerks are required to solve specific problems. Same-sex couples were one example and, each time one was identified, a clerk was alerted to revisit the original forms. Various checks were then made to confirm that the apparent same-sex couple was 'genuine'. Other information on the original forms helped in the identification of errors made by the respondents – the most obvious being that the wrong box had been ticked on the form. Thus, some individuals that were blood-related, rather than partners as the form suggested, were identified. However, errors could also have been introduced by the ONS. In certain cases where information was missing from the returned forms the ONS imputed data and, occasionally, an individual's sex would have been missing from the original form. The sex of some individuals could have been imputed incorrectly, creating fallacious same-sex couples and this would also be checked by the clerk. Of course, after these thorough checks a number of genuine same-sex couples would have remained. The clerk was then instructed to alter the coding in such a way that the householder's partner was recoded as an 'unrelated' member of the household. In Table 4.1, this would mean culling those gay individuals who described their situation as, say 'husband/wife' or 'living together' and placing them instead into the 'unrelated' category. Thus, the exclusion of same-sex couples from the census output resulted from this *conscious manipulation of the census data*. Allegedly, of course, the rationale for the change was not so much a direct exercise of heterosexism or homophobia *per se*, but rather a scientific concern with reliability. Here it seems the governmentality of the closet operated through the epistemology of science that underpins the state's *savoir* of its population 'at a distance'.[22] More precisely, the concern that drove the alteration of data was scientifically justifiable in terms of reliability: the criterion that each time the question was posed, it measured the same thing.

The census offices did at least make some attempt to quantify the number of couples that were being recoded in the 1991 returns. During this editing process, the opportunity was taken to count those cases where a clear statement of same-sex cohabiting status had been made. A brief analysis of this information was carried out during the processing period and based on five-sixths of the 10 per cent Britain sample only 360 such couples could be identified.[23] This information was published in a short article in *Population Trends*, which concluded that 'the results bear out the earlier indication from the 1989 census test that the 1991 census question on relationship could not properly identify gay or lesbian couples' (Office of Population Censuses and Surveys 1993: 1). Since the census question was not designed to capture such couples this may not be too surprising. Interestingly, the decision has been made that this minor manipulation of census information will not be carried out in the 2001 census, although the wording of the question on household relationships that has been tested to date makes no mention of same-sex couples at all! While this argument was used at the time, it was probably the difficulty and cost of altering the computer programs that was the more forceful reason.

It appears that the closet was produced through heteronormativity at the irreconcilability of two scientific aims, rather than some direct policy of homophobia. On the one hand, data were recoded in order to preserve the principle of reliability. The explicit changing of the data, however, raises the worrisome and equally important issue of validity: that the measurement device (in this instance the question) actually captures what it is it is supposed to measure (arrangements within the household). Governmentality seems to have been exercised through this impasse, producing a closet that conceals lesbians and gay men in the nation.

## Conclusion

In this chapter we have moved from the urban scale to the national scale. It has spatialised the closet at the national scale in the United States and Britain. While gay and lesbian studies have examined the closet through a number of framings of the nation (from the literary to the judicial to the military), Foucault's concept of governmentality was introduced as a means to expand that operationalisation. The term describes the network of power relations between the state and population that operate through the categories through which the former knows the latter. We suggested that the census is an interesting example of the governmentality of the closet. Using the US and British censuses, we tried to 'see' gays and lesbians in the two nation-states. Neither country asks a specific question about sexuality or the nature of same-sex households. In this broad sense there is certainly a closet governing the way gay people cannot be counted directly through this portrait of the population. The story is more complex than that, however. In the US the number of same-sex couples that could be identified also suggests a certain closet governmentality. Here we noted the complexities of governmentality. The tyranny of categories also meant, for instance, that only gay people in hierarchically organised couples could be identified. Likewise, stressing the self-governance aspect of governmentality, respondents' concerns over privacy and homophobia could certainly explain why more people did not describe their relationships in such a way that we could have allowed their identification.

The point of these exercises certainly was not to cast aspersions or lay blame for the closet at the feet of individual respondents or the census departments. Indeed, the interesting point is just how multifaceted the forces that produced the closet were. In the cases above the closet results from structures of heterosexism that conjoin with issues over personal privacy, administrative efficiency and scientific rigour.

Here the ethical question arises: should gays and lesbians be counted on the census? If they were, the closet would partially be dismantled at this spatial scale.[24] Gays and lesbians would arguably increase their visibility in the nation given this additional opportunity to 'come out'. It would be compelling evidence that they are, in fact, everywhere. The statistical legitimacy of this incontestable *fact* would be powerful, indeed. Given the legislative importance of the census, the impacts for congressional representatives might be quite direct. On the other hand, arguments could certainly be made against self-declaring one's sexuality on a government form. Despite the assurances of confidentiality, worries over privacy certainly do still exist. People would, in effect, be forced by law to admit their sexuality, since it is illegal to lie on a census form. What would happen, for instance, if PUMS data revealed an otherwise hidden enclave of lesbians in a politically conservative census track? What kind of politics might result from that knowledge being publicly available? Before one can take an ethical stance on the question, we would press them to consider the variety of interweaving sources of power/knowledge that already place the closet in the census. Simply arguing for categories that better reflect gay and lesbian relationships on the form, for instance, would not necessarily dismantle the self-imposed closet, for instance – though it would certainly be a start!

While we stand behind our arguments about the relationship between the closet, governmentality and the census, we recognise the dangers of over-extending the point of this chapter too. For example, one might argue that lesbians and gays do not use the census principally as a means of framing them-selves nationally. In this way, other modes of representation become 'more important' to interrogate vis-à-vis the closet. One commentator, for example, noted the ability of marketing firms or insurance companies to 'see' gays in the nation through their own profiling techniques and statistical prowess. We cer-tainly agree that the census is not the only possible framing for the closet. Like-wise, we definitely do not want to be interpreted as saying that the census is the only location of the national closet within the state itself. The vignettes that opened this chapter, and the charged debates over 'family values', age of consent laws, same-sex marriage, immigration policy, inter alia, obviously demonstrate the census is one – albeit important – part of a broader constellation of state power/knowledge around sexuality. In sum, we see our contribution as rather modest: to point out how the census is a location for the closet in the nation in both simple and complex ways. We see this contribution as relevant because this dimension of the census has been neglected by queer studies population geog-raphers and students of governmentality to date and is a particularly interesting example of Foucault's ideas.

This chapter has shown the relevance of seeing the closet as a material force at the national scale. When we think about the presence of the closet in the nation, we would do well to think beyond the likes of Peter Mandelson, Ron Davies, Bowers v. Hardwick or 'don't ask, don't tell'. The closet appears in far less sen-sational national representations. It exists in the census forms and data files through which the population is defined and governed. Moreover, as an exercise of power/knowledge, its operation is at once simple and complex. There is no question on the form about sexuality, to be sure. Yet gays and lesbians can still be seen if they, and the people looking for them, make a variety of assumptions. Ultimately, the national closet is not a simple space, nor is it simply metaphorical. It is, as queer theory predicts, a knowing by not knowing, owing to the vagaries of heterosexist categories and bureaucratic rationalities. Nonetheless, the closet is also a not-knowing by knowing as well. Closet governmentality is sustained not merely by surveillance techniques but by its assumptive relations with other epistemologies we have.

# Notes

1 School of Geography and Geosciences, University of St. Andrews, St. Andrews, Scotland, UK.

2 Even after his fracas Davies neither discussed his presence on the Common nor declared his sexual orientation publicly. He did not stand for the Labour party in a recent election of candidates, though he is still an MP (but not a cabinet minister).

3 When asked in an ICC/*The Guardian* poll, 'Do you think homosexuality is morally acceptable or not?', 56 per cent say yes; 36 per cent say no; 7 per cent don't know (see Travis 1998: 1). Note that this is a general question, not directed at whether they think homosexuality is acceptable for an MP specifically.

4 Brown (1995) has explored the issue of distance in scientific epistemology.

5 Of course this point is also a fiction of the state. We know that censuses are never fully comprehensive – in the 1991 British census approximately 2.2 per cent of the population failed to complete the form: the 'missing million'. In the US debate is raging over enumeration methods on both statistical and political grounds.

6 Note that the variable gets extensive use in American applied and academic research.

7 These are Census Tracts numbers 74 and 76, which are bisected by Broadway E., the heart of Capitol Hill. A rather similar exercise was undertaken by Castells (1983) to reveal the Castro and other gay neighbourhoods in San Francisco. Forsyth (1997) offers a more recent example.

8 In a classic essay on queer theory, Scott (1990) has argued against the epistemological weight of experience (specifically in historical research), noting that this seemingly direct evidence is never theoretically unmediated. Here we are trying to make the inverse point: supposedly neutral, scientifically legitimated and governmentally endorsed 'facts' in the census produce a closet on their own, unless we draw on other epistemologies. It is not only experiential or phenomenological knowledges that are mediated!

9 Even if details on this category were provided, most would not be same-sex couples. Further data processing would be needed to provide output on same-sex couples from these data.

10 Micro-level individual data *can* be purchased by anyone. However, they are certainly much less accessible to the general public in terms of time and money, than the tract-level output.

11 The investigators are Paul Boyle, Tom Cooke, Keith Halfacree and Darren Smith (see Boyle *et al.* 1999).

12 The geographical identifier is far more detailed in the US than in Britain.

13 The head of household is called the 'householder' in the US but both are referred to as head of household here.

14 The debate over the number of gays and lesbians is a fascinating topic itself, worthy of an analysis of governmentality. Much of the debate turns on methodological issues, and questions of how one actually operationalises determinations of sexuality (behaviour, sexual practices, self-proclamation, personal histories, etc.).

15 This argument needs qualification. We have only extracted couples that include a householder because of this problem. In other words, this problem of self-identification of a hierarchy within the household is operating in the data across both straight and gay couples.

16 Again, this would also 'closet' many straight couples (like academics!) who live apart perhaps because of distant work locations.

17 We could also identify the PUMAs (Public Use Micro data Area) where these individuals live. PUMAs offer us relatively detailed geographies of localities but with only 138 individuals, it is impossible to make any sense of this distribution.

18 Census divisions are groupings of states that are subdivisions of the four census regions. There are nine divisions, which the Census Bureau adopted in 1910 for the presentation of data. The regions, divisions and their constituent states are:

**Northeast Region**
*New England Division:*
Maine, New Hampshire, Vermont, Massachusetts, Rhode Island, Connecticut
*Middle Atlantic Division:*
New York, New Jersey, Pennsylvania
**Midwest Region**
*East North Central Division:*
Ohio, Indiana, Illinois, Michigan, Wisconsin
*West North Central Division:*
Minnesota, Iowa, Missouri, North Dakota, South Dakota, Nebraska, Kansas
**South Region**
*South Atlantic Division:*
Delaware, Maryland, District of Columbia, Virginia, West Virginia, North Carolina, South Carolina, Georgia, Florida
*East South Central Division:*
Kentucky, Tennessee, Alabama, Mississippi
*West South Central Division:*
Arkansas, Louisiana, Oklahoma, Texas
**West Region**
*Mountain Division:*
Montana, Idaho, Wyoming, Colorado, New Mexico, Arizona, Utah, Nevada
*Pacific Division:*
Washington, Oregon, California, Alaska, Hawaii

19 Two individual-level data sets were provided for the first time in the 1991 British census. The individual-level SAR is a 2 per cent sample of the total population, while the household-level SAR is a 1 per cent sample of households. All of the census variables are attached to the individuals in these samples, although in the household SAR individuals are linked within households. This is the sample used here.

20 Note that one popular reason for the 'missing million' was the large number of people who had refused to pay the infamous 'poll tax' introduced by the Thatcher government. Presumably, many of these people were unwilling to risk being traced through the census information, and were suspicious of the confidentiality claims. This may also have been the case for lesbians and gays.

21 This office is now called the Office for National Statistics (ONS), as identified earlier in the text.

22 This is not to imply that there was not a more general structure of heteronormativity at work. For instance, the very fact that it was only recently that this question came up suggests a wider context of closeting. Our point here, however, is to study the

micro-practices of power, and so we concentrate on the manifestation of closet space through scientific rationality.

23 Each household completes the same form, but due to expense, some questions are only coded for 10 per cent of the returns (detailed occupational status, commuting to work information, etc.). The SAR data are drawn from this 10 per cent sample.

24 Interestingly, a sort of closet would remain for people in heterosexual relationships who have gay partners. The categories still fail to capture the complexities of people's lives.

# 5

# WORLDING THE CLOSET

## Desire, travel and writing

### Where in the world is . . . ?!

'Too chicken to come out? You closet cases expect me to weep for you, to cry over your hard and tortured lives. But no can do –I think you're a bunch of pathetic losers,' asserted *Out Magazine* columnist Dan Savage (1999: 34) recently. His polemic went on to list the reasons why people all over the world tell him they stay in the closet. Of course the one that caught my eye was number five: *location*. The argument he dismissed was that it is more difficult to be 'out' in certain places than others. To Savage's mind, however, this reason was just as specious as all the rest and summarily rejected:[1]

> This defense has gained currency in the wake of the murder of Matthew Shepard. But if it's impossibly or wildly dangerous to be out where you live –and people get bashed in New York City, too, by the way –then consider moving. What's more important to you, living honestly or living where you do right now?
>
> (Savage 1999: 34)

Now there are several problems with Savage's retort to be sure, but I find two particularly bothersome. His view of the closet is thoroughly metaphorical, lacking any potential for the materiality of power/knowledge oppressions it names. He deploys the closet in his argument as a simple metaphor for the hiding of one's homosexuality, what I identified in Chapter 1 as the comparison theory of the closet. Furthermore, he presumes a queer body that has no spatiality whatsoever (or at least none of any social consequence). It is an autonomous body located anywhere on a Cartesian plane. The issue, we know from poststructural theories, is that metaphors cannot be *just* simple rhetorical flourishes: they draw on and inform social theories and ways of worldmaking. In this way Savage's conceptualisation of the closet is one that emphasises *being* over *desiring*, and neither is influenced by either's situatedness.

Savage's oversimplified argument raises the question of how we might spatialise the closet in the world. Any mode of representation that attempts such a global perspective is inevitably partial, of course, yet several possibilities come to mind in the form of texts. World political geographies like Kidron and Segal's (1984) or Amnesty International's (1994) not only locate us 'everywhere', but also detail the patterns of nation-state laws against queers and their sexual acts. There is also a substantial market for international lesbian and gay travel, and it is evinced by advertisements, guidebooks, and the like (Gmunder 1998; Van Gelder 1991). These books tell us where (and when) to go to be out. They tell us places to avoid when we are out. More recently, the internet has globalised sexual desire in a paradoxical way.[2] Cruisingforsex.com, for example, lists maps and describes locations for anonymous sexual encounters in public space across the entire planet![3] We can now download erotic images of bodies from around the world. We can even have cybersex: sex without intimate proximity. In a way, these texts are a means by which desire and placement exist in global space because of – and in direct resistance to – the closet. The texts that I wish to consider here, however, are those of gay travel writing. I am moved to consider this empirical venue in part because it is a burgeoning yet largely unexamined literature (Rist 1993; Ebensten 1993; Miller 1989; 1992). Examination of gay travellers and their writing, in fact, has largely been historical, and certainly not oriented centrally around the spatialisation of the closet (Burg 1983; Aldrich 1993; Bleys 1995; Gregory 1995; Phillips, 1999).

By examining travel writing, I want to move beyond the frustrations thrown up by Savage by thinking particularly about its relation to gay desire around the world. I will do this by prompting what Said (1983) and Clifford (1992) would call 'traveling theory'. They agree that theory's inevitable mobility (across disciplines, between material locations, and amidst different theorists) carries us to the frontiers and boundaries of theory's ability to explain. It is not surprising, then, that travel writing – or more accurately the geographies it reveals – might be used to confront the utilities and the limits of a particularly influential body of theory. I would like to goad such a travelling theory in this chapter, namely psychoanalytic accounts of desire, and more specifically their own travels into literary theory. I will argue that these accounts can be brought into conversation with written geographies to produce critical insights about the mechanics of desire. I want to think through some of these relations by considering the travel writing of American gay author Neil Miller. In his two books, *In Search of Gay America* (1989) and *Out in the World* (1992), Miller explores (quite literally) the geographies of the closet at national and global scales. This penultimate chapter will focus on his travels across the globe in which he navigates the closet. He describes similarities and differences that highlight the range of ways the closet can work on desire. I will draw on two often-competing foci on desire in psychoanalytic/literary theory: Lacan's and the schizoanalysis of Deleuze and Guattari to analyse the workings of the metaphor. Miller's travel writing shows that where we desire enables and constrains how we desire.

## Psychoanalysis and literary theory

The links between psychoanalytic theory and literary criticism are tangled and not always clear. And at first glance, the two intellectual pursuits seem interested in entirely different fields: the psychoanalyst is interested in the human unconscious; the literary critic is interested in the written text. There has been a substantial exchange of ideas between the two camps though, as Brooks (1987) points out, it has been largely one-sided, with literary theory being largely the recipient of insights from psychoanalysis. The rationale for theoretical travel between these two domains can be understood through three closely related claims. First, scholars claim that the two enquiries have similar subject matter. Here we see Lacan's (1977: 147) famous claim that 'the unconscious is structured like a language'. Gallop (1985), for instance, argues the association is based upon an analogy: the literary critic is like the psychoanalysts. Likewise Wright (1984) advises critics to work on a text as a psychoanalyst would work on a psyche. Thus the parallels in the production of signs' meanings through metaphor and metonym can be drawn between a person's dreams and a piece of literature.

It follows, then, that if the subject matter is similar, so too are the tactics of investigation, and this is a second connection. Both literary criticism and psychoanalysis rely on the premise that there is a hidden level of meaning below the surface. What either a text or a human behaviour seems to mean at first glance is not necessarily what it does mean. In psychoanalysis, Freud's 'discovery' of the unconscious established this occluded world of signification. In literary criticism, Eagleton (1983) argues, the (often implicit) notion of a 'subtext' presumes exactly the same structure of meaning. Both pursuits therefore devise strategies of deepening or expanding their reading of things. Both search for contradictory, or ironic, truths that stand in tension to the more 'conscious' or 'literal' meanings at hand.

Finally, both their subject matters and their techniques illustrate what Selden and Widdowson (1993: 136) call 'the articulation of sexuality in language'. Questions about our relationship with desire, and the signification systems that we use to express or suppress it, are clearly bound up in both pursuits. So, for instance, Eagleton (1983) suggests that there is a confluence between the two intellectual pursuits because both are interested in the same thing: how human beings pursue pleasure. Similarly, Brooks (1987: 4) claims their paths cross 'where literature and life converge'.

Psychoanalytic literary criticism, then, seems to be an appropriate starting point in considering the relations between desire, text and space when it is spatialised globally through travel writing.[4] It enables an understanding of the way the closet metaphor works textually in travel writing. Conversely, travel writing about the closet, like Miller's, would speak to the utility and insightfulness of this mode of criticism.

## Miller's geographies of the closet

Neil Miller is not a scholar *per se*; he is a journalist by trade. He was editor of Boston's *Gay Community News* in the mid-1970s and more recently has written for *The Boston Phoenix*, an entertainment and arts weekly. Consequently, his self-professed aim with these travelogues was quite descriptive. As he put it, 'I didn't start out with any particular preconceptions, any grand theory I wanted to prove. My intention was to paint up-close portraits of people and communities, letting my subjects tell their own stories' (Miller 1989: xv). Following the points raised in the earlier section, these books do have theories travelling through them: theories about space and desire. The impetus for the books came in the mid-1980s, when he noticed the extent and speed with which AIDS had forced people out of the closet, a process Altman (1988) has referred to poignantly as 'legitimation through disaster'.[5] The AIDS crisis propelled people out of the closet at a number of scales. It forced many people living with AIDS to disclose their sexuality publicly. As it hit gay men through the 1980s, their communities became much more visible as they fashioned responses to the epidemic and demanded others take notice of the holocaust in their midst (Brown 1994). Miller watched these trends avidly. He witnessed a 'moving in from the margins' unfold in his native Boston, as well as in other urban gay centres like New York, London and San Francisco (Miller 1989: 9). The simple comparison closet metaphor seemed to be failing him in these urban texts. Gays and lesbians were becoming more visible; they were gaining an explicitly, widely recognised presence in places. Yet simultaneously, he admitted, it was as apt as ever. 'Some things just didn't change.' A rise in anti-gay violence could be documented by the mid-1980s. It was still considered acceptable to be anti-gay in public discourse. And even a 'liberal' place like Massachusetts explicitly barred lesbians and gays from being foster parents. The paradox was enough to send Miller packing. He decided to investigate this duality and document it through travel. He toured small towns across the United States. In these locales he sought out lesbians and gay men to talk about their lives, their positions in and out of the closet.

> I wanted to know, 20 years after the Stonewall Riots if gay pride and progress had finally begun to trickle down to the grassroots; if the

options for living, working, and being belligerently ourselves that we had won in cities like San Francisco, New York, and Boston extended to the rest of the country, to the towns where so many of us had grown up, to minority communities, to mid-size cities. I wondered too, as a gay man living in a comfortable urban environment, if I might have something to learn – about survival, about community and finding one's own place in society – from people in less congenial places.

(Miller 1989: 11)

A few years later he undertook a more global expedition of lesbian and gay life, again to chronicle its existence inside and out of the closet. While his books are by no means comprehensive or even entirely representative of gay and lesbian life – and Miller acknowledges that partiality explicitly in his writing – they nonetheless provide a certain in-depth yet extensive cartography of gay and lesbian identities in places. They show how sexual identity and desire are geographically mediated. And in this way, they inevitably situate the closet in global space.

Miller's conclusions at the end of his travels perhaps come as no surprise to any postmodern human geographer. For while there is surely a certain degree of 'outness' in the willingness to talk to a journalist, the people he meets in a variety of different locations tell complex spatial stories about heteronormativity and homophobia. Across these narratives Miller finds no common theme, no patterned geography of desire, no model architecture of the closet. In his own words, 'I found no single vision of gay future. Instead I saw various gay and lesbian communities taking different paths, adopting their own strategies as they attempted to create an atmosphere of openness and a sense of security' (Miller 1989: 306). As substantively interesting as his travel writing is, I think the plurality it tries to represent has important theoretical implications for psychoanalytic theories of desire in literary work, and the metaphor of the closet in particular. Through his work, I find myself tracking theories' (in this instance, of desire) travels between the material closets inhabited by desiring lesbians and gays and the metaphorical closets of desire built through psychoanalysis. Miller's travel writing is nothing less than a global geography of the closet. It charts where and how gay and lesbian desire takes place. In order to trace this route, it is necessary to sketch out the ways that psychoanalysis has theorised desire itself.

## Geographies of desire

While there are many geographies to take from Miller's work, I am most struck with the way it mediates psychoanalytic theories of desire spatially.[6] I will specifically focus on the rift between Jaques Lacan's (1977; 1994) Oedipally[7] centred theory of desire and Deleuze and Guattari's (1983) 'schizophrenic' account. Both of these long-standing perspectives continue to inform psychoanalysis' take on how desire works, and their insights on that process also have 'travelled' towards other disciplines, like literary theory (Eagleton 1983; Wright 1984; Selden and Widdowson 1993).[8]

To stake my argument on a difference in thinking between Lacan and Deleuze and Guattari is not to presume they disagreed *completely*. Indeed, as Sarup (1989) sees it, Deleuze and Guattari sought to politicise key insights of Lacanian theory, which they felt had revolutionary potential. As poststructuralists, both Lacan and Deleuze and Guattari reject a unified theory of the subject.[9] Each accepts a necessarily decentred subjectivity and the argument that subjectivity and desire are tightly bound up with one another. Both aspire to portray desire much more diffusely than simply placing it all upon a *specific* object of desire. Bogue (1989: 3) argues more specifically that Deleuze had long been hostile to the implicit Hegelianism in Lacan's theory of desire.[10] He first critiqued it in *Nietzsche and Philosophy* (1962). *Anti-Oedipus* (Deleuze and Guattari 1983) is a widening of that rift with Lacan, albeit an explicitly politicised one given Guattari's activism. This seminal book is a trenchant critique of the Freudian insistence on the Oedipal origins and mouldings of human desire. While Lacan, too, draws away from Freud in important respects (see Bowie 1979; Grosz 1990), his work clearly extends and reinforces an Oedipally centred human desire. Both are also trying to resist Freud's more biological explanations for human desire that pervade psychoanalysis, especially those that source desire in some innate instinct. This is especially true around the Oedipal complex, the supposed stage of sexual development in children around the ages of 5 and 6, where a child's sexual desires are sorted out by gender (see note 7). While Lacan rejected Freud's arguments that there was something fundamentally teleological to this 'stage', Lacan – and especially his followers – nevertheless accepted its ontology as a stage, and explained it more precisely through the child's ever-deepening immersion into social structure and culture (see below).

Despite these shared beginnings, however, the two theorists have developed ideas about desire that are quite at odds with one another when they have travelled into literary theory. Typically, they are portrayed as poles in a debate about how desire works (e.g. Hocquenghem 1993; Bogue 1989; Selden and Widdowson 1993; Grosz 1990; Fuery 1995). And therefore, they lead us into rather different readings of texts on desire.

### *Lacan and desire*

Geographers such as Gregory (1997) and Pile (1996) have teased out the thoroughly spatial epistemologies at work in Lacanian thought. Here, however, I want simply to concentrate on Lacan's notion of desire as fundamentally a lack. For Lacan, desire and subjectivity are inseparable (Fuery 1995). His meditations on desire come from his quest to understand how subjectivity forms in the infant. When we desire, we are trying to assert a sense of being. It is a way to mark our existence. Yet with his conceptual orders or 'registers' of the Real, the Imaginary and the Symbolic, he charts the formation of subjectivity by metaphorically locating subjectivity not only corporeally but also simultaneously in broader social structures and processes (Gregory 1997). In this way he moves far from the centred, corporeal and essentialised theories of subjectivity so characteristic of Modernist thinkers (Sarup 1989). The Real is the seemingly natural, a priori order into which the child is born and through which it just begins to self-identify as existing. It is a seemingly complete and simple space that Gregory (1997: 210) characterises as 'the lack of a lack'. Leader and Groves (1995: 22) explain the Imaginary this way:

> Mastery of one's motor functions and an entry into the human world of space and movement is thus at the price of a fundamental alienation. Lacan calls the register in which this identification takes place 'the imaginary', emphasizing the importance of the visual field and the specular relations which underlies the child's captivation in the image.

The Symbolic is the 'outside' world of culture and language. It is omnipresent and inescapable as the infant works out its sense of identity and placement in the world. By entering the symbolic, Lacan argues, the subject wants to possess the signifier from which he or she is always alienated (the phallus). But because of the Oedipus conflict, and our attempts to resolve it through the mirror phase, desire can never actually be satisfied. It can never actually be vanquished because of its otherness for (and through) social subjects. Lacan theorises the child in a quest to understand and become what the mother wishes and desires, which he labels the phallus. The phallus is thus an imaginary lack that is always beyond the child, yet something that the child recognises s/he is at least part of. Lacan insists that we explore the phallus through language, and the child's entry into that structure.

Desire is 'other' according to Lacan because it shows subjects that their needs cannot be self-satisfied they must look elsewhere, outside the conscious self.

> The subject, driven by desire, enters the mirror stage and subsequently the Symbolic in search of a satisfaction that can never be achieved. This is 'desire' understood as an active process. Once the subject enters the Symbolic, finds that he or she is alienated from the signifier (that is, from all systems of representation and articulation, especially language), and

becomes split within himself or herself, desire becomes even more pronounced and problematic.

(Fuery 1995: 17)

It is perhaps less important that we wade through Lacan's wilfully dense rhetoric and logic than focus on his portrayal of desire itself. In the Lacanian view, desire is always a *lack* for the subject. In their accessible introduction to his work Leader and Groves (1995: 83) explain his thinking well: 'If demand is demand for an object, desire has nothing as its object: nothing in the sense of "lack taken as an object".' Desire is never satisfied; it never can be, but is constantly deferred. It always exceeds discrete desired objects themselves. Thus desire is a metonym, a sort of horizontal substitution of signifiers (Lacan 1977: 167). This is why he makes the well-known claim that the unconscious is structured like language. Given that subjectivity and desire are so closely bound up with one another for Lacan, we are said to be defined by our desire. In other words subjects are constituted by what they lack – not through any whole or unified subjectivity (which is the conventional model of humanity in western philosophy). Uncertainty, fragmentation and alienation therefore mark desiring subjects.

The account above charts how the closet can be written in travel writing. It frames our view of the closet as an empty, hollow space, a negative-space of being in the world where desire is defined as an absence that leaves our sense of self incomplete. Recall that I am interested in the allegedly spatial qualities that allow the closet to work as a metaphor. Lacan is certain that subjectivity is defined by a lack: by *uncertainty* (a lack of full knowledge), *fragmentation* (a lack of complete-ness or cohesion) and *alienation* (a lack of belonging). These features also high-light certain material aspects of the closet in Miller's geography. Indeed they resonate with closets identified in previous chapters of this book. Miller's writing situates the closet by showing how they matter in people's lives, as well as how they matter intellectually to our ideas about desire.

Let me begin with simple examples of the closet as a spatial manifestation of a lack of full knowledge, a lack of completeness or cohesion, and a lack of belonging. In his characterisation of life in the South African closet (cf. Elder 1998), even in its post-apartheid manifestation, Miller shows us how segregation materialises alienation:

> For many years, there had been gay communities in cities like Johan-nesburg and Cape town, but they had been quite closeted – and totally white. White gays could enjoy the advantages of bars, bathhouses, a monthly newspaper, even a pivotal role in a race for a Johannesburg parliamentary seat. But for blacks, there had been virtually nothing. Until a few years ago, blacks hadn't been allowed to visit black townships

without a permit. Apartheid had kept white and black homosexuals from the most basic social interactions, let alone the sense of a unified community.

(Miller 1992: 8)

We can read a similar alienation between race and sexuality that closets black lesbians and gay men in South Africa in Miller's conversation with Phybia, a black lesbian who explains her frustration with the alienation the closet manifests:

> Nonetheless, there was a sense of excitement among gays and lesbians in South Africa, and a visitor couldn't help but feel it. The world was opening up. Phybia, the drama student who had driven out to kwaThema with me, had had her picture on the cover of a Norwegian gay magazine. She participated in the gay and lesbian march and was involved in the newly formed gay student group at her university. Only a few months before, she had been closeted and traumatized about her attraction to other women. 'Because you're black, you're not expected to be gay,' she told me. 'People say, "It's white people who are gay. You're not gay. You're African. It's not part of your culture." But look at me. I'm here!'
>
> (Ibid.: 9)

Consider for, example, this narrative of Tokyo, where the closet is materialised in the city:

> For the visitor, particularly the non-Japanese-speaking visitor, Ni-chome was a closed world. The bars were almost impossible to find. They were located along lengthy corridors in what appeared from the outside to be small office buildings; sometimes there were three or four bars to a floor. There were virtually no street signs in Tokyo and, even where there were, street numbers signified nothing. The neon signs that advertised the bars were invariably in Japanese. Someone gave me directions to GB, a bar popular with *gaijin* (Westerners): 'Turn left at the fruit stand a couple of blocks past the coffee shop, go down the alley, and make a left and go down the steps to the basement. But if you go on Sunday, the coffee shop will be shuttered and you will get completely lost.
>
> (Ibid.: 146)

His writing on Japan also speaks to the theme of separation and fragmentation around a cultural distinction between public and private spheres. As in other places, a tightly circumscribed private sphere instantiated a closet where same-sex desire could take place. He likens Japanese culture to a series of compartments or rooms, with far more rigid walls between public and private spheres than his own American culture. He tries to appreciate the low permeability between *tatemae* (social convention) and one's inner feelings (*honne*), and locates the closet here for gay Japanese.

On this separation and fragmentation, Miller compares Japan and Thailand. In both places, he sensed that 'same-sex relations never implied a gay identity'. Here we might read a profound alienation in the lack of a wider supportive community in such places. Connections between subjects are made only on the lines of desire; not friendship, politics, nor solidarity. He realises this point in a discussion with Kanda, a twentysomething high school English teacher who lived in the US and was active in a gay-youth organisation:

> 'In the United States, "coming out" suggests the formation of new kinds of personal relationships with heterosexuals. But in Japan, coming out won't accomplish that. Heterosexuals just don't understand us or care about us.' What gays needed to do, in his view, was to strengthen themselves personally and as a community and to forget about the rest of society.
>
> (Ibid.: 156)

Likewise, his conversations with a variety of Japanese gays and lesbians revealed the profound alienation between queers and straights. Hirano, a Japanese teacher in Tokyo, explained it to Miller this way, '"In Tokyo," he said, "gay people will say they are gay in a gay context only. Otherwise, they stay in the closet"' (175–6). This alienation he explained by the hegemony of marriage, which was so relentless, pressed on by family, friends and co-workers.

> It was unclear which created more problems – to be gay or to refuse to marry. 'The sole obstacle Japanese gays would face in coming out is the marriage system,' as one gay man put it. 'You can live with things as they are, as long as you clear up the question of marriage with family and bosses.
>
> (Ibid.: 163)

Even if one can manage to overcome that level of alienation, there is yet another – between the omnipotent, traditional family in Japanese society and the infrastructureless same-sex lovers, or *koibito*, that some lesbians and gays are trying to forge. Minako explained it to him this way:

> 'It is like taking hostages,' she continued. 'People use your ailing mother or your ailing father to keep you in place. We always have to choose

between respecting our ties with our original family and our new relationship. It is always one or the other. It is real psychological pressure and
you get mixed up.'

(Ibid.: 169)

There was also alienation between gay men and lesbians because of the closet.

As long as gay men remained closeted and pretended to be hetero-
sexuals, they received the privileges of the dominant gender. Only by
coming out, renouncing their privileges as men, and allying with women
could gay men transform themselves and society.

(Ibid.: 176)

Turning to his travels in the non-metropolitan United States, Miller insightfully plays on Lacanian themes of desire in his travelogue. While Selma, Alabama,
is a medium-sized American city, its location in the Deep South makes it a
culturally conservative and generally homophobic place. Gay and lesbian desire
there is clearly structured by the closet. Listen, for instance, to how Miller
describes what the closet is like for 'Jill', a 40-year-old white lesbian he met in
Selma, Alabama:

Like other gay people she knew in Selma, Jill was in the closet. None of
her neighbors were aware that she was a lesbian, except for one woman
who was her best friend. She hadn't told anyone at her job, and she never
set foot in the gay bar in Montgomery, out of fear of exposure.

(Ibid.: 18–19)

There were no gay bars in Selma, no gay, lesbian, or feminist groups, and
no gathering place except Skeeters, the bar in Montgomery 58 miles
away. 'I'm not sure there could ever be a place here in Selma,' said Jill.
'Maybe it's paranoia. But I think it would not be tolerated.'

(Ibid.: 18)

She went on to say: 'It's not even like the gay grapevine here because people are
so far back in the closet, they are back behind several rows of clothes. Maybe
underneath the shoes' (ibid.: 18).

The themes of uncertainty, fragmentation and alienation are also manifest in
Miller's discussion of Hong Kong in the early 1990s, before decolonisation.
Officially speaking, homosexuality does not exist in mainland communist China.
Same-sex desire is depicted as a western corruption. Yet at the time of Miller's
trip, nineteenth-century anti-sodomy laws were still on the books. Cultural and

legal structures have combined to construct a closet that strives to discipline, deny and erase same-sex desire. Thus in the colony, the closet works in several ways to promote the discourse of lack around desire. Most obviously, Miller describes how it was impossible for him to meet any lesbians in Hong Kong. They were too closeted for him to reach. As well, he goes on at length about how the closet in Hong Kong perpetuates the absence of a strong gay and lesbian community. He describes how gays and lesbians are starved of any sort of gay culture because of the ubiquity of the closet, and suggests this lack leads to a profound sense of uncertainty about who they are. And perhaps rather ironically, he discusses how gay and lesbian desire within the closet leads to a dangerous lack of anonymity. The need to preserve one's anonymity in Hong Kong, Miller shows, leads to a profoundly fragmented and alienated gay or lesbian subject. As one of his informants confided:

> Your friends are all around you. Your colleagues are very close to you. If something is known, the news passes quickly. Mostly, people are afraid of their sexual orientation being known, so they try not to go to the disco or the bar. Outsiders know these places. So you meet others at the public toilet, instead. A quickie. But what about mental need? The spiritual need? You just keep it inside your heart. Closet! That's the word.
>
> (Ibid.: 100)

Later, another informant, a writer who was daringly open about his sexuality, went on to describe just how powerful an effect the closet had on the relationships between people's identity and their desires:

> . . . he viewed contemporary Hong Kong as a big closet. 'We are geographically blocked', he said. 'There is simply nowhere to be anonymous. You can't drive to Shanghai like you could in the 1930's.' Some of the symptoms of the closet were letters he had received after his books were published. One reader wrote . . . using his left hand, so his script couldn't be identified. Another letter was composed of Chinese characters cut out of newspaper, for the same reason.
>
> (Ibid.: 104)

From these passages we can heuristically link up the physical space of the closet with Lacan's account of desire. A closet is typically separate from the room it adjoins. It is at the margins of immediate space. In the places Miller describes above, the closet perpetuates a lack around desire through tropes of alienation, uncertainty and fragmentation. It is kept separate from the immediate lifeworld.

It is kept at bay not only from the public sphere, but from the private sphere as well. To be in a closet physically is to stand apart from, but still inside, the room where the closet is located. It is to be alone. In both Hirano in Tokyo and Jill in Selma the alienation gays and lesbians feel is intense, even where gays know one another. Jill says they are 'back behind the shoes'. For Lacan, alienation is also a symptom of desire, which can never be sated. Standing apart from others in an actual closet, one cannot always be sure what goes on in the room itself. It is hard to see the centre from the margins. And even though hooks (1984) reminds us that a marginal location can be a site of empowerment, her insights do not deny the already well-recognised power imbalance between the centre and margin. In Selma, Jill reflects that she might be paranoid, fearing how straights might react to same-sex desire in Selma. But she can never really be sure. For Lacan, uncertainty always surrounds the desiring subject. We can read uncertainty, fragmentation and alienation in the imagery of the actual closet, just as we can in the intersection of subjectivity and desire of gays and lesbians in the different places Miller describes.

### Deleuze and Guattari and desire

Reading the closet as I have above seems to be at odds with a Deleuzian theory of desire. And yet such an alternative framework produces just as plausible and helpful reading/rendering of the closet. Deleuze and Guattari have provided a strong critique of both Lacan's and Freud's followers. Their central argument against these early psychoanalysts homes in on the fact that far from being a neutral, innocent tool for the investigation of the psyche, psychoanalysis is itself thoroughly part and parcel of the structures and discourses of power that can harm the self. They are especially angered by the reductionism through abstraction that flows directly from the interpretive function of psychoanalysis. The notion of desire as based in lack or need, which implies some sort of essential deficiency in human subjectivity, can only ever be recognised and cured by that which theorised it in the first place. They decidedly reject the Oedipus complex as the panhistorical, ubiquitous law that defines all human desire in terms of a deficit. So they provocatively ask:

> . . . is there an equivalence between the productions of the unconscious and this invariant – between the desiring machines and the Oedipal structure? Or rather, does not the invariant merely express the history of a long mistake, throughout all its variations and modalities; the strain of endless repression? What we are calling into question is the frantic Oedipalization to which psychoanalysis devotes itself, practically and theoretically, with the combined resources of image and structure. And

128

despite some fine books by certain disciples of Lacan, we wonder if Lacan's thought really goes in this direction.

(Deleuze and Guattari 1972: 53)

That representation of desire Deleuze and Guattari attribute as a function of capitalism. It is a tactic of interpellative, hegemonic control at the most micro-scale. Writing amidst the heady French (post)structuralism of the late 1960s, they insist that the mode of production 'deforms' the unconscious, a rather provocative extension of Freud's thinking about the social structuring of human consciousness. But parting with Freud, they argue that psychoanalysis cannot cure but only exacerbates things. It denies the complexity of lived reality, and furthermore is a means of disciplining and confining – thus ultimately repressing – desire through interpretive structures and the hegemonic authority of psychoanalytic practice. The Oedipus complex, they insist, is a result of capitalism's general repression and channelling of a rather effusive human desire. Capitalist social relations, working through the institution of the modern family, repress desire. While a quarter of a century later we might be suspicious of the reductionism their own work exudes, they nonetheless stand as tireless critics of the standard theories of desire, not to mention purveyors of an alternative way of thinking about it.

Their alternative approach is to celebrate the potential for a multiplicity of desires. They concatenate the signs of 'desire', 'machine' and 'production' in an attempt to alter the way we think about ourselves as agents who desire. They label their perspective 'schizoanalysis', where desire is not lacking but productive for the subject. They identify with the schizophrenic form of desire because in it they see revolutionary potential in that condition's refusal to conform to pre-given structures and assumptions, and its often clever and ingenious reworking of signs and meanings. There is no single, correct form or direction desire will take. It is to be thought of as 'an untrammeled flow' (Selden and Widdowson 1993: 143) in order to liberate the subject from the oppressions of social norms. They deploy the metaphor of the 'desiring-machine' in an attempt to deterritorialise and decentre desire from a specific form of human development, as well as link desires more explicitly with the world beyond the body. From this iconoclastic premise, then, there is no such thing as misplaced desire (which the Oedipal theory assumes). In turn, desire need not signal any necessary lack or deficit in the human subject. 'Desire isn't a lack,' they would say, 'it's just desire.' Rather than being contained by Oedipal anxiety, desire is conceptualised as an energy, a positive source for new beginnings, and a 'voyage of discovery'. Rethinking desire in this way, Deleuze and Guattari call it productive, creative, generative and multiple. Through schizoanalysis, they seek to liberate desire from the limits and stigma placed on it by capitalism, and exacerbated by modern psychoanalysis. They call this strategy 'deterritorializa-tion', suggesting motility instead of fixity. They want to liberate desire from oppressive institutions (or 'territories') of the family, church, nation, school and, of course, psychoanalysis itself. And following on from the notion of a decentred subject, their notion of a desiring machine locates desire not simply in the body of a subject or in another object or body. Rather, the metaphor asks us to think about subjects being located within networks that produce desire for and through us. In terms of conceptualising the closet, schizoanalysis deterritorialises desire from the previous reading offered above. It asks us to consider the ways gays and lesbians *do* attempt to desire in particular contexts – especially though in spite of the fact that such contexts may impel them to conceal or deny their identity or desires. More broadly, it also asks us to consider the cost of only seeing the oppression of the closet (as Savage does), rather than the cunning, heroic resistances to it. In the context of literary theory, schizoanalysis asks us to read desire in this new and productive way in texts (Selden and Widdowson 1993).

Deleuze and Guattari's perspective has had some appeal in queer theory. Hocquenghem (1972) for instance has drawn extensively – albeit problematic-ally – from their work to expose the biological essentialism and heterosexism that saturates both Freud's and ultimately Lacan's work on desire.[11] He uses schizo-analysis specifically to criticise Lacan, whom he positions as largely a Freudian progeny because of his uncritical adoption of the Oedipus take on desire. As Weeks explains (1978: 30),

> The major problem is that Lacan, like Freud, appears to make these stages, and the Oedipal Complex a transhistorical human experience, though for Lacan it is essentially a cultural not a biological experience. Even Juliet Mitchell, who believes the Law of the Father can be eventu-ally overcome, believes it to be a necessary element in patriarchal societies.
>
> For Deleuze and Guattari, as for Lacan, the forms of desire are not set in nature but are socially created. But they reject psychoanalysis, and in doing so construct a challenge to Oedipus as a *necessary* stage in human development. They attack Lacan for staying within the Freudian family framework: as a result, psychoanalysis is trapped *within* capitalist economic and social demands (emphases original).

Hocquenghem (like many queer thinkers and activists influenced by the sexual revolution) stresses the important *political* point that gays and lesbians should not necessarily import definitions of desire from an essentially heteronormative psychoanalytic framework, but work creatively to produce their own (see Calfia

1995). Like Irigary's feminist critique, Hocquenghem shows that the stakes of a singular Lacanian reading of queer desire are high indeed! He explicitly challenges the heteronormativity that, by reading desire as a lack, so easily marginalises same-sex desire and reproduces a system of beliefs predicated on the fact that there must be a centre, a normalcy to desire. Ultimately, then, psychoanalysis cannot help queer-desiring subjects, it can only alienate and marginalise them by pathologising their desire.

> The homosexual perversion must submit to the rule which assigns certain objects to certain drives in an exclusive way, just as it must submit to the rule of fixation to the parental person: these bonds are needed to stop the drift of desire.
>
> (Hocquenghem 1993: 118)

> Desire, as an autonomous and polymorphous force, must disappear: in the eyes of the psychoanalytical institution, it must exist only as a lack, or absence. It must always signify something, always relate to an object which will then become meaningful within the Oedipal triangulation. This is now the position of post-Freudian psychoanalysis, which is an institution of bourgeois society charged with controlling the libido.
>
> (Ibid.: 77)

Hocquenghem takes the marginal, deviant position of homosexuality in both French society and psychoanalysis seriously and then pushes the analysis to its inevitable conclusion:

> Homosexuality is thus defined as a lack. It is no longer one of the accidental specifications of polyvocal desire, but is assumed to signify hatred of woman, who is the only social sexual object. Heterosexuality is 'full', as opposed to a homosexuality which lacks the essential object of desire.
>
> (Ibid.)

> Homosexual neurosis is the backlash to the threat which homosexual desire poses for Oedipal reproduction. Homosexual desire is the ungenerating-ungenerated terror of the family, because it produces itself without reproducing.
>
> (Ibid.: 107)

Reading Miller's travels with this more generative perspective on desire provides a rather different – but no less accurate – geography of the closet. I want to be clear about what I am suggesting here: I am not using schizoanalysis to argue that there were smaller or larger closets in different places. Nor am I saying that there were more people in the closet in some places Miller visited than others. It is simply that when I think about the geographies of the closet I find a Lacanian reading of desire seems to deflect attention away from the cunning, generative strategies lesbians and gays have used to desire – in spite of, or in opposition to, the closet. This is a powerful motif in Miller's writing because he is at pains to witness the variety of ways people in different places around the globe are managing to desire people of the same sex in spite of (or through) the closet. In this way I would argue he attempts a schizophrenic literary account of the gay desiring-machine.

Indeed, it starts off his book. Recounting a trip to Israel in his youth, he described a closeted encounter that reads not like a lack, but as a strategy of a wilful, generative homosexual desiring-machine:

> During the two years I spent in Jerusalem, teaching English, in the early 1970s, I made the acquaintance of only one other gay man. We met on a bus from Tel Aviv. The bus was crowded and we both stood in the back, gripping the same railing and trying to avoid being swept down the aisle. As the bus climbed the Judean hills, I felt the slightest pressure of his fingers against mine. I didn't pay any attention. Then I felt the pressure again. We didn't look at each other or speak. We were both in the closet in a closeted society.
>
> (Miller 1992: xiii)

Returning to Hong Kong, despite the thorough ubiquity and intensity of the closet, Miller admits that the closet did, in fact, have a geography: the bar, the disco and the public lavatory. In other words, there were spaces productively carved out within the closet where desire could be pursued in spite of alienation. More obliquely in Selma, despite the fact that there was no discrete gay space in the city, people could still make contact with others who shared their same-sex desires.

> Even in a town like Selma gays have an uncanny ability to make contact. 'You know the expression, it takes one to know one?' asked Jill. 'You just *know*, and once other gay people realize you are gay, they will make comments and you get to talking and they tell you things.' But all this was terribly discreet.
>
> (Miller 1989: 18)

Jill herself had recently been romantically involved with a woman from Birmingham, and while she would often commute there, the couple would often spend the weekend at Jill's house in Selma, her neighbours wholly oblivious to the situation.

Still another strategy was to stand within the hetero/homosexual dualism, to be 'inside/out' in a sense. Whether this could be called bisexuality or 'feigning' heterosexuality is debatable. The point is that several people Miller met were in different-sex marriages and conducted same-sex affairs as well. Miller documents the pervasiveness of this strategy in several locations about which he writes. For example, he describes a chance encounter with a young gay man in Hong Kong:

> As we started back down the hill, he began talking rapidly and excitedly. He had been living a double life, he said, trying to hide from his family, his friends, and, above all, from his girlfriend. At work, he made sure to avoid gay people, who might have seen him near the bar or the disco; he feared that talking to them might give him away. He couldn't introduce his gay friends to his girlfriend. 'What would I say? This is my Canadian friend. This is my New Zealand friend.' What would she think?
>
> (Miller 1992: 106)

Likewise, Miller recounts his meeting with Mr Hassan in Egypt. Mr Hassan was a teacher who lived in a suburb of Cairo. He would travel to the West – the United States specifically – he would come out.

> Mr. Hassan led an active sex life, much of it revolving around quickie sex in hotel rooms and a downtown cinema (now shut). He told me that just the previous week he had met an Italian man at the Hilton health club and gone to his hotel room, along with two other men.
>
> (Ibid.: 83)

The sharp divide between inside and outside the closet in Japanese culture is also used productively according to Miller's account:

> For many Japanese gays and lesbians that meant there was no urgency to come out and to tell family or friends about one's sexual orientation; it also meant there was no need to avoid marriage. You could divide your life. You could have sex behind a screen in a public place, and people were supposed to pretend it wasn't happening, as long as when the screen was removed you were sitting there perfectly composed with your tie on straight.
>
> (Ibid.: 149)

And in his account of gay life in Buenos Aires, Argentina, during the Proceso, Miller deftly juxtaposes both perspectives on desire as he explores the closet there, one enforced by a violent state apparatus:

> People coped in a variety of ways. Some gay men married women. Others tried to appear as straight as possible. 'You imitated traditionally masculine characteristics,' one man told me. 'It was common for gay men to involve themselves in sport.'
>
> (Ibid.: 197)

Miller goes on to describe the rise of literary societies, café-concerts, dinner parties, and a complex, frequently changing schedule of cruising areas. The diversity of strategies for meeting, organising and articulating desire in a single place surely locate schizoanalysis in his text.

Hocquenghem decried the source of latent Oedipalised homophobia in capitalism's need for social reproduction through the family unit. The case of Thailand's sex-industry shows that the desiring-machine he argued for gay sexuality is no less immune from capitalism's functionalist incorporations of sexuality. There, Miller notes how the district of Patpong capitalises on productive desire.

> The bars – gay and straight – that cater to Western sex tourists are mostly located between Silom and Surawongese roads, in a several-block area known as Patpong, after the street of the same name. Patpong is the most international part of Bangkok, a fact that says much about Thailand's relationship with the West. Airlines like Quantas and Swissair have their offices there, and a number of expensive hotels, shops, and restaurants are within walking distance.
>
> (Ibid.: 116)

He goes on to read the displaced colonising desires of Western travellers to Thailand not so much as part of the local desiring-machine, but as capitalism's transformation of it:

> Sex tourism was colonizing Thailand in a way the European powers never had. Still, the boys seemed to rise above it all somehow, like those crowds in their perfectly ironed shirts crawling out of buses in the ninety-five degree heat. Hospitality? I doubted it. More likely, just commerce.
>
> (Ibid.:118)

Though this was not the only story to be told about Thailand. Leaving the heart of the colonising sex-trade, he travelled to Sukumvit Road and located gay bars that catered to locals specifically. Unlike its garish colonial counterpart, it was closeted and full of closeted men, largely. Thai men did not simply automatically place their desires in the local economy of the sex industry; they created their own alternative spaces of desire. Miller recognises this point when he visits a closeted gay bar for locals called 'Moonstruck', located on a small *soi* (side street), at some distance from Patpong:

> Ninety percent of his [Mr Wut's] clients were Thai, he said with the remainder Taiwanese and Japanese. Almost no Westerners came to Moonstruck; they preferred Patpong's mix of souvenir shopping and entertainment, along with sex. Here, his customers were interested in sex, and sex only. Out of fear of being seen, many Thais wouldn't go to Patpong; even at Moonstruck, they would often wait outside in the car. 'Thai people lose face if they go into a gay bar,' Mr. Wut said. 'They are ashamed.' Many of these customers were, in fact, married men. 'Deep in his mind he is gay, but he must marry,' said Mr. Wut. 'This kind of person may wait in the car, even let the captain choose a boy for him.'
>
> (Ibid.: 120)

In other words, just because people were (desiring) in the closet did not mean that they did not resist and adapt to the situation. Desire produced strategies of resistance because of the closet.[12] A much more expansive map of desire unfolds in *this* picture of the closet too.

The very presence of his informants – and Miller's own ability to meet and network through them – in these places suggests a certain degree of resistance emerging from the productive, generative aspects of desire. A consistent theme in Miller's work is how creative gay and lesbian people have become in pursuing their sexual desires in the context of a homophobic closet. This suggests a certain productive dimension to closeted desire that may ultimately act back upon the constraints it imposes.

Now clearly I have not discussed the obvious constraints that limit these people's ability to pursue their desire creatively. Nor have I discussed the ethics or dangers of their chosen strategies. But that is because my point is not to argue the merits of these tactics. It is instead to offer a different geographic reading of desire, one that is informed by Deleuze and Guattari's schizoanalysis. In a closeted place (or a place of closets?) like Hong Kong or Buenos Aires desire was being 'deterritorialised'. It was, as Miller discovered, transgressing the limits and fixity placed upon it by the family, the church and the community. Gays and lesbians were actively liberating their desire. Recall that schizoanalysis portrays desire as productive, creative, generative and multiple. Each of these qualities might be used to describe the interior of the closet in Miller's travel writing. People devised several interesting and successful ways to pursue desire. Miller himself seems to recognise the power of schizoanalysis when he reflects on his travels through the closeted small towns of America. The variability, strength and successes of their people awe him:

> What struck me once again was this movement towards a richer, fuller life. . . . You couldn't have a one-dimensional existence if you lived in this part of the country. . . . Gays in small towns and rural areas had traditionally had those community ties and involvement. Usually they had been closeted – often in the deepest closet of all, a heterosexual marriage. Now many were cautiously trying to combine an allegiance to their roots with some of the openness and community that gays had in big cities. And so many people I met in these smaller towns seemed to have gone through enormous changes within the last six or seven years. There was still a fear of exposure, still homophobia of course. AIDS, while seemingly remote, heightened prejudice and suspicion but also enhanced the possibility of forging links with other gay people.
>
> (Ibid.: 105)

It seems to me that we risk missing an important political point if we only take a Lacanian reading of desire when we spatialise desire through 'the closet'.

The final geography lesson that Miller imparts back to Lacan and psychoanalytic theory itself gives rise to broader questions about travelling theory, those raised by postcolonialism. In fact, the relations between queer theory and postcolonialism have only just begun to be considered, as scholars recognise the global diffusion of queer culture from the West (Altman 1996; 1997) and try to understand resistances to same-sex desire in other cultures (Hilson 1996). So it is important to recognise here the politics of this travelling theory: psychoanalysis is certainly a colonial discourse itself, powerfully overgeneralising and essentialising not merely on the identities of gender or class, but also of race, time and space. Lacan's is a European theory of the desiring subject, one that desires in the middle of the twentieth century. Indeed, this point was raised early on by the authors of schizoanalysis: 'Oedipus is always colonization pursued by other means, it is the interior colony, and we shall see that even here at home, where we Europeans are concerned, it is our intimate colonial education' (Deleuze and Guattari 1972: 170).

The colonial implications of classic psychoanalytic theory of desire are perhaps best evinced by Miller's account of Egypt, where he is forced to recognise the thoroughly western orientation of the category 'homosexual'. While there were no gay clubs, organisations, or bars there was the *hammam*, a Turkish bath. It was a place where men could go to have sex with men. It was a closeted place in familiar ways. Miller describes it as dark and decrepit, for instance. But it was here where he realised same-sex desire was deterritorialised out of his own western categories. In Egypt he found such desire relatively commonplace, yet sharply sanctioned. Male-to-male sex was never discussed, and only those who had spent considerable time in the West self-identified as 'homosexual' or 'gay'. For others, the terms did not describe same-sex acts or desire. They were seen as foreign impositions, even if they were used locally to describe desiring subjects. There he found himself questioning the validity of the categories he used to inform his very quest: gay and homosexual. From such cultural awareness, his journey meant he was constantly negotiating between the universals and particulars of being queer in the world:

> Just what did 'gay' or 'lesbian' or even 'homosexual' mean? Was a gay or lesbian identity (as distinguished from behavior) merely a Western cultural notion? How relevant was the dichotomy we set up in the West between 'homosexual' and 'heterosexual', between 'gay' and 'straight' to a non-Western society, say Egypt or Thailand? Was it a mistake to hold up Western-style openly gay or lesbian models for such countries?
>
> (Miller 1992: xv)

Now to be sure, Deleuze and Guattari's own European roots can hardly be denied. Moreover, Miller is self-admittedly an American travel writer: his own power/knowledge grids cannot escape the colonising gaze either. Nevertheless, both texts are willing to recognise the myriad forms that desire – as a productive, creative response to the closet – can take. On this point, then, schizoanalysis forms a rather more democratic and less imposing means of appreciating desire. What such a reading of Miller's work cannot address, however, is whether or not the closet itself is a form of colonisation: yet another export in the diffusion of queer culture from the West. Does speaking of the closet make sense everywhere, beyond the West? If it does, is that because of a globalising economy and culture that export it?

## Conclusion

The purpose of this chapter has been to use travel writing to produce a travelling theory. I have used psychoanalytic theory's migration into literary criticism to unpack the spatial metaphor of the closet. It suggests that the closet is also a spatialisation – a set of material spatial arrangements – that are integral to gay and lesbian desire. Different takes on how to theorise desire, however, have produced different insights into the workings of the closet in a particular opus of travel writing. Lacanian theory points us towards the function of concealment at work in the metaphor. Located within the closet, lesbian and gay desire can be difficult for anyone – gay or straight (even the subjects themselves!) – to find. Alternatively, schizoanalysis draws our attentions towards the productive capacities of desire precisely because of their confinement as deviant and out of place. The closet does not necessarily alienate people from desire, or necessarily prevent them from translating it into wants, demands, or wishes. I would like to close this essay by drawing out two implications for psychoanalytic literary criticism when the closet is spatialised, as Miller's travel writing has done so effectively.

First, Miller's in-depth/expansive geographies of lesbian and gay life paint a thoroughly equivocal picture of the closet for psychoanalytic theories of desire. It can be coded through a Lacanian take on desire, where the closet is a space of lack. It is a place of constant deferment and never-sated desire. It is often a place where people are striving to make themselves more complete but will always fail. But it can also be coded as a tremendously creative and productive space of desire. A schizoanalytic reading of the closet is just as plausible. Closets in material space were bursting with ingenious and productive strategies of desire. I get a much stronger sense of the equivocation of desire in Miller's work than I do from psychoanalytic texts. So more specifically what I wish to argue is that for all its poststructural rejection of dualisms and categorical imperatives, maybe psychoanalysis has some important geography lessons to learn about bothness, duality and betweenness in placing desire and describing its workings. By approaching desire through the closets in Miller's travels, we can see how a positioned academic debate occludes the simultaneity opposing perspectives can share in place. Returning to Said and Clifford's notion of 'traveling theory', I suggest that it is not so much that theories travel, but that theories are always travelling *and* staying fixed in place.

Second, I would close by reflecting on what we should make of this closet of desire. Criticising it as too static (as a number of geographers have suggested for spatial metaphors more generally) risks downplaying the vital political point that fixity is precisely how the closet can conceal and trap gays and lesbians (Brown 1996). The static nature of the metaphor points up a significant oppression. But on the other hand stressing the limitations the closet places on desire draws attention away from an equally important – if antithetical – political point: that lesbians and gays in the closet can and do desire successfully through a number of creative and generative strategies.

One might argue that, as a geographer, I have fundamentally misread psychoanalytic/literary theory. That may be true; but if I have, I think the point about travelling theory which opened this chapter becomes all the more salient. Travelling theory is no longer done just by closed-off academic communities (of, say, literary critics or psychoanalysts). Consequently, when such discourses adopt spatial metaphors, they must be willing to import geographic theories about the dynamics of space and place.

I think the most useful way to approach the closet, then, is as a 'translation term' between sexuality, space and desire. As Clifford (1992: 110) explains, this is 'a word of apparently general application used for comparison in a strategic way'. He hastens to add to this definition, 'all translations used in global comparisons . . . get us some distance *and* fall apart' (emphasis original). The closet gets us (and the theories we travel through) some distance in understanding the confinement and erasure of lesbian and gay desire, but it collapses when we approach the creative and innovative productions of desire people have pursued in places/ closets simultaneously.

## Notes

1 The other 'excuses' he listed were: 'mommy and daddy', 'job', 'I'm not just gay, I'm so much more complex', and 'Emmy winners need to come out [before I can]'.

2 Far from rendering geographic theory quaintly outdated, cyberspace is a quickly growing and fascinating ambit for geographers, precisely because it simplifies and complexifies spatiality. For general perspectives see Crang, Crang and May 1999. On gay and lesbian's use of cyberspace see Brown 1995 and Wincapaw (forthcoming).

3 This knowledge need not necessarily be extra-local. For example, I discovered from this website that a men's room in the building in which I work was a favourite cruising area on campus – and that it was being subject to surveillance!

4 The relationship between psychoanalysis and queer subjects, however, has been by no means unproblematic (e.g. Bayer 1981; Lewes 1988). One might therefore view this essay as an attempt by queer theory to speak back to psychoanalysis.

5 See Altman (1996) where he makes the argument about the role of the AIDS pandemic in helping to diffuse a global queer culture.

6 For instance, his travels demonstrate what Luke (1994) has termed 'glocal' politics, as struggles to create livable local gay space can have effects at a variety of other spatial scales. It also strikes me that Miller's travel writing has much to inspire the recent spate of comparative work in urban politics, and attempts to sensitise them to scale issues. Another important geography Miller self-consciously explores is the negotiated one between lesbians and gay men. He is convinced that a viable politics of sexuality must include both gays and lesbians. Thus this travel writing pays close attention to the coalition-building work done to overcome the so-called 'limits' of identity politics that have often split sexual politics along gender lines.

7 Briefly stated, the Oedipus complex is a standard masculinist theoretical construct in Freudian psychoanalysis that describes a normal emotional crisis early in a child's psychological development as a heterosexual being. A boy's sexual impulses towards his mother and jealousy of his father characterise this stage. The consequent guilt that develops as the child begins to understand the social relationships into which he has been born signals the rise of the superego through castration anxiety. For girls, the path is more complex. Girls are said to develop penis envy in lieu of the castration complex, exemplified in hostility directed at the mother for her 'lack'.

8 Psychoanalysis, of course, has also travelled into geography, and several geographers

are interested in a return trip. For some recent examples see Pile 1996; Pile and Thrift 1995; Rose 1993; Bondi 1993; Blum and Nast 1996; Gregory 1997.

9 On the debate about whether Lacan can be 'properly' called a poststructuralist, see Leader and Groves 1995: 78. By labelling him as such, I am signalling his acknowledgement of structures (the autonomy of the symbolic) being paralleled by his insistence on retaining a place for the subject.

10 Fuery (1995: 14–15) notes that the specific influence is in Lacan's use of the master–slave model from Hegel. See also Sarup 1989: 21–2.

11 The problematic nature of this text lies in its masculinism. I draw on Hocquenghem's specific argument nonetheless, because it makes an early and progressive intervention into the heteronormative biases of psychoanalysis generally.

12 This is perhaps Chauncey's (1994) point in arguing against the metaphor of the closet in favour of the metaphor of a 'gay world' to describe New York in the early twentieth century. See Brown 1996.

# 6

# CONCLUSION
## Where is the closet?

## The argument

My aim in this book has been to ruminate on the possibility that the closet is not *just* a metaphor for the concealment, erasure, or ignorance of gays' sexualities. Though it works in several metaphoric ways, there are other ways to approach this sign theoretically. It can have spatiality, an existence in space that has location and situation, which signifies placement, interaction, movement and accessibility. That spatiality matters to its very nature (that nature being a power/knowledge of homophobia, heterosexism and/or heteronormativity). By its spatiality the closet is a material strategy and tactic: one that conceals, erases and makes gay people invisible and unknown. Because it is such a common, central term in gay and lesbian life, it implies a ubiquity and multidimensionality that suggest an exploration across a wide variety of spatial scales and locations. Theoretical work on the closet's epistemology concurs, suggesting its complexity: a 'knowing by not knowing', a 'deadly elasticity'. Here I took Sedgwick's claim that the closet's presence has had both deep and widespread cultural effects to heart and argued that conceptualising the closet geographically meant it spoke to a wide array of social thought: from performativity to urban theory, to governmentality and to psychoanalytic theory. It simultaneously presented itself at several spatial scales from the body, to the city, to the nation, and finally to the globe.

## The geographies

Chapter 1 introduced the concept of the closet through prevailing theories of metaphor: comparison, interaction and poststructuralism. Each suggested important trajectories of meaning for the metaphor. Comparison theory asks us to relate homosexuality to a particular kind of space. Interaction theory builds on that comparison and suggests that in our meanings of sexuality we sort out ways the closet both is and is not akin to being gay or lesbian in a heteronormative

141

society. Poststructuralism presents us with the vexing truth that there can be no *retrait* from metaphor. It furthermore acknowledges the possibility for the renegotiation of meaning of any sign as it travels through time and space. Each of these accounts, however, skirts the possibility that the closet is not always just a means of representation. It does not simply just name something else. So this book has asked, 'What if we acknowledged that the power/knowledge of the closet was itself always-already a spatial process?' This move was meant to draw on and extend geographers' recent attempts to reject the dualism between metaphor and materiality when discussing space and spatial metaphors.

The closet's spatiality must be considered in ways beyond the meaning of the literal architectural sense, of course. Thus I have attempted to locate the power/knowledge of homophobia, heterosexism and heteronormativity in a variety of locations around the globe. Furthermore, I have tried to follow my colleagues in geography who are grappling with spatial scale and its inevitable politics (e.g. Bell and Valentine 1996; Herod 1997): to recognise that the closet works through, and can be materialised 'in', scales of the body, the city, the nation and the world. To rethink the closet through what Gregory (1994) would call 'geographical imaginations' begins with an axiom often unknown outside the discipline: that all social relations are fundamentally spatial (just as they are temporal). From this point we can appreciate how space configures heteronormative/homophobic structures and agencies in ways that may be both simple and complex.

Chapter 2 located the closet at the scale of the body. It was a body with a voice, which spoke about his coming out. It was textualised through a series of coming-out narratives of American and British twentieth-century gay men. These men brought a rich geographical imagination to the closet: one that spoke to the currently popular theories of performativity. Performativity asks us to think of gender and sexuality like language, where one does something by saying it. It asks us to notice the 'scripts' that are always-already provided for us when we try to perform our gender and our sexuality. The script must be already there if we want to improvise as well. Trying to speak back to its linguistic roots, I asked performativity to notice the complex relations between speech, silence and the closet. Saying can surely be a performative coming out of the closet – but so too can it be a means to remain inside. Silence can be a means of staying in the closet, but ironically it might also be a means to leave that metaphorical space. In recent scholarship, geographers have been working assiduously to underscore the importance of space to performativity. I tried to extend that work by noticing how the closet's spatiality could exist in multiple ways: a shy introverted self, a hidden beat in the city, or a place from which to emigrate.

The axiomatic argument that all social relations are spatial originated with Lefebvre's writings in the late 1960s and early 1970s. Revolutionary as his thinking has been to consider the force of abstract space in knowing the city, it generally remains mired in heterosexism and essentialist assumptions when it discusses sexuality and urban space. To reorient and recuperate his work, I drew on a textual analysis of the landscape of sex and desire in Christchurch, New Zealand, a place where sexuality and desire have come to define the inner city. The landscape of highly visible straight massage parlours and saunas affirmed his arguments about the imbrication of capitalism and abstract space that shatters women's bodies into fragmented commodities. Considering gay sexuality as part of the production of urban space makes it clear that the closet is materialised in urban space. Hidden, concealed bars and sex-on-site venues are the very opposite traces and marks of heterosexualised space. Yet both take place in exactly the same part of the city. Reconciling these two spatial strategies was attempted by extending Knopp's (1992) emphasis on sexuality's relationship to capitalism as a means of understanding urban space and introducing Harvey's (1996) notion of the spatial fix. We might think of them together as flexible-yet-fixed arrangements of urban space that produce a (visual) landscape, and attempt to solve crises that inhibit capital accumulation. Here the closet is not *only* an effect of heterosexism or homophobia; it is an effect of the imperative of accumulation as well.

In Chapter 4 Paul Boyle and I placed the closet in the form of the census at the scale of the nation. Looking at the census as a textual representation of the nation-state, we conversed with Foucault's notion of governmentality. It is the 'mentality of government' both in the sense of how scholars of politics think about government's power and how the state itself thinks about the object of its power. Foucault's history of the modern state identifies a shift away from sovereign, top-down exercises of power like brute force on subjects by police of military. In modernity the state comes to rule in large degree over a population by always knowing it expertly at a distance through bureaucratic, scientific and statistical epistemologies. Governance is achieved by a state working to solve the problems of a population. And the integral part of this way of knowing is national censuses. They describe and document and thereby create a population of a nation. The knowledge they produce is legitimised through both their reliability and their validity, and work at a micro-level of power through their self-reporting organisation. Taking both a macro and micro approach with census data, we explored the extent to which we could 'see' lesbians and gays through the census in the United States and the United Kingdom. The closet manifests itself in both simple and complex ways through this power/knowledge of the census. Simply put, there is no question about sexuality on either census, and that fact is clearly a closet space at the national scale. It is also more complex than that, however, because categories that might capture such an identity do so only if assumptive links are made with other epistemologies of the closet (e.g. partnerships, the ecological assumption of a queer neighbourhood, shared residence, etc.). We demonstrated this point by looking at Capitol Hill in Seattle, an ostensibly gay neighbourhood in the US. Certain gays and lesbians could be seen, but only by closeting others. Thus, for example, we could see same-sex couples (and even that required an assumption that cohabitants were more than room-mates) but ultimately we would end up closeting queers who were single, polygamous, or not living with their partners. These epistemologies circulate not only amongst readers and researchers of the census, but respondents too! Turning to the UK micro-data, we found an interesting case of heteronormativity in the recoding of same-sex couples from 'husband/wife' and 'living together' to the 'unrelated' category. Here it seems as though the scientific reliability undergirding the census' epistemology of the nation binds to heterocentric assumptions about the nature of relationships and the result is a closet, albeit one that consequently challenges the validity of the epistemology of the census.

Placing the closet at the global scale was Neil Miller's travel writing, which was discussed in Chapter 5. This richly descriptive, empirically grounded work portrayed the complex negotiations gays and lesbians make in a wide variety of contexts. Miller's writing provided an especially geographic problematisation of Lacanian psychoanalytic theory. Though he differs with Freud in important ways, Lacan reproduces the Oedipal metanarrative that operationalises desire as an essential lack. To a degree, Miller's travels do spatialise the closet as lack. He shows us the closet as isolation, fragmentation and uncertainty in a wide variety of geographic settings. He shows us how certain places do seem to contain closets that isolate gays and lesbians from the ones they desire, as witnessed by Miller's depiction of places like Buenos Aires, Selma, Thailand, Japan or Hong Kong. He found fragmentation through the lack of solidarity amongst queers in countries like South Africa and Japan. Miller portrays the struggles of gays and lesbians in these places as struggles to desire.

Drawing on Deleuze and Guattari's schizoanalysis – which is itself critical of the essentialism of the Oedipal metanarrative in psychoanalysis generally – Hocquenghem critiques such a perspective, calling it heterosexist and homophobic. If desire is essentially a lack, queer desire is one that can never be fulfilled, and is misogynistic to boot. Instead, Hocquenghem argues we should think of desire as decentred, fragmented, but also adaptive, productive and creative. It is a generative, polymorphous and multiple force. Understanding desire in this way shows the closet to be a rather different strategy of power/knowledge. It is one that is structured to allow queer desire to occur in a heterosexist space. It is part of a strategic resistance, however compromised, to homophobic regimes in various places across the globe. Miller offers us a simultaneous, yet contrasting, geography of the closet in the same places and elsewhere. And it is this latter reading he is at pains to stress in his travel writing. Miller emphasises the myriad and ingenious ways gays and lesbians articulate and act on their desires while placed in the closet. Such strategies are as simple as a gentle touching on a crowded bus in Israel. It also includes the more ineffable 'gaydar' Jill tries to explain to him in Selma, or the dual life of people in South Africa and Japan. Even the very absence of an indigenous category of homosexuality in places like Egypt, Miller ponders as a possible manifestation of the desiring-machine. Miller's travel writing shows us that the closet's cultural geography is often freighted with contrasting theories of desire. However one reads these multiple and variegated desires, his travel writing is nevertheless self-admittedly and inescapably singular in its perspective: a western, middle-class American gay man. His positionality raises the spectre of the globalisation of gay culture as a new form of colonisation. More modestly, it raises the question of whether or not the closet is essentially a western space of desire, and if it is, how then do we engage in a radical-democratic queer politics of resistance to 'it'? How can we see it globally without colonising through our travelling theories of desire, sexuality and space?

These empirical chapters collectively speak to a need within queer theory and studies to understand the closet not just as a metaphor for something else. If we agree that the closet 'stands for' a diverse array of exercises of power/knowledge deployed to conceal, erase, or deny queers subjects, we must admit the possibility that those exercises themselves can be quite spatial indeed. The fact that we can see different manifestations of the closet in different places makes it no less spatial. What is more, there are times when those spatial tactics are quite blatant and simple: the careful heterosexualised performative, the hidden gay bar or cruising spots. True enough, other times these tactics are rather complex: the family members who 'always knew', the subtle codes in public space, the accurate yet imprecise category of 'unmarried partner' on a census form, for instance. Yet if we rush to queer theory's present epistemology of the closet as a 'knowing by not knowing' *alone* we risk missing the quite explicit and direct ignorances that the closet materialises in social space. We must find a way to theorise the simplicity and complexity of the closet. The re-placement of the closet at a variety of spatial scales, to my mind, opens up a series of theoretical insights that speak to thinkers and literatures across the humanities and social sciences. The fugue of this text, then, was not just scalar; it was intellectual as well.

## The limitations

There are a number of limitations to this study that I acknowledge and revisit here in the spirit of honest reflexivity, partiality and the hope of spurring future scholarship. Foremost is the itinerant closeting of lesbians and other sexual minorities by a nearly exclusive focus on gay men's closets. Several authors both within geography (Chouinard and Grant 1996) and outside it have argued that the trend of queer scholarship to focus solely on gay men's lives is tantamount to erasing the presence of lesbians (Jefferies 1994; see also Jagose 1996). To be fair, I would point out that I have tried to include lesbians where I felt it was not co-optive to 'speak for the other'. For instance, Chapter 4 demonstrates the way governmentality may closet both lesbians and gay men. Additionally, Miller's travel writing in Chapter 5 recounts his discussions with both gay men and lesbians, in part because of the unity and solidarity he witnessed between them in certain places. In guiding my politics here, I have tried to remain faithful to Alcoff's (1991) argument. While it is often problematic to explain or describe the oppression one does not personally inhabit, we cannot deny that there are also times when it is truly irresponsible not to bear witness to them.

The inevitable use (implicit or explicitly) of binaries like inside/out or gay/ straight is surely a limitation in the context of a plethora of works whose aim has been to deconstruct and blur these binaries and dualisms. They can imply a static fixity that is not necessarily there. They can prioritise the first term in the couplet to the marginalisation of the second. On the other hand, as Derrida himself pointed out, to deconstruct a dualism does not mean to reject it entirely, and rejecting one set of dualisms inevitably means invoking another. My point here is not to be disingenuously ignorant of the messiness our common terms obfuscate, but rather to use terms in ways that – at this stage of our intellectual conversations – I do think are politically useful and theoretically important to the extent that they awaken queer theory to its neglect of spatiality.

The singularity of my own perspectives on the closet also limits the generalisability of this research. While I have tried to see the closet from as many different perspectives as possible across the lives of different people, undoubtedly my own situatedness enabled me to see some and not others. Similarly, my own theoretical and intellectual curiosities meant I conversed about the closet with a narrow set of theoretical debates (and even then only on certain terms). These were the closets I could see, and undoubtedly there are others to chart – yes, even within the 'texts' I have examined. The conversation, I hope, should not end here. Foremost (especially after the discussions in Chapter 5), the thoroughly western epistemology of the closet needs to be interrogated specifically, though I hope in a sense of empathy and radical democracy.

Finally, an especially vexing – and ultimately unsolved – problem with this book has been to find a textual strategy that melds the simultaneity of scales to the closet in a single place. In the end, I am left feeling the chapters' rather simple and boxy architecture fails to convey the interrelatedness of spatial scale that is invigorating other debates within geography. I have organised this book as a search for the closet in a variety of locations and at a diversity of scales. I did this because I think it is a politically important step (even if it is an awkward one) to see the similarity and continuity of homophobia and heteronormativity that ontologically *is* 'the closet', despite the considerable range of differences in size, shape, intensity and effect.

## The implications

There are at least three lessons that I would proffer from this research, in spite of its limitations. First and most directly, I would insist on the value of a geographic perspective for queer theory. It is valuable to understand the spatiality of the closet because it is important to understand the different-yet-similar material ways power works to oppress – just as it is important to see the different ways queers attempt to resist! The closets I have named so spatially might be boringly familiar to most gays and lesbians: the introverted self, the lie, the phoney marriage, the nondescript city bar, the park at night, the census form or data, etc. Others have charted their existence over time. My point here was not to 'discover' them but to rethink them. Whatever else they are, however else we might understand them, they are spatial, and power/knowledge works through them spatially *inter alia*.

The second implication I would raise is a refusal to explain the closet as *either* a simple or a complex manifestation of spatial power/knowledge. It is both. The complicated, paradoxical and confusing dimensions of power that the closet names have been insightfully characterised by queer theory: a knowing by not knowing, a coming out that is also a closeting. Even simple aphorisms like 'coming out is a continual process' affirm the complexity at stake, recognising that complexity need not mean rejecting spatiality, however. Indeed, what I have tried to do is expand on these insights through the case studies. People can be in and out of the closet simultaneously through resituating that broader environment. Its space can reveal and conceal at the same time, often dependent on one's own location. Mobility and migration can materialise coming out, but they might also be a way of simply coping from within a larger closet. Forcing some people out of the closet can ironically closet others.

Rather paradoxically, however, I don't want to surrender the signification of fixed isolation and entrapment upon which the metaphor trades either. Indeed, I would hasten to remind my geographic colleagues that in our rush to criticise the static spatial metaphors abounding in cultural studies, we should not forget the very real fixity of absolute space, and its utility in explaining oppression. Certain spaces *do* conceal, erase and deny the existence of marginalised groups, and they do so in quite simple and straightforward ways. People can put themselves there, or be forced there. Sometimes it just is that simple. I want to think about the closet with all the theories of metaphor from Chapter 1, because that's the (multiple) truth of it. As soon as I admit that certain aspects of homophobia are not at all like the space of the closet, I have to remind myself that presuming forever-fixed relation between signifier and sign is a convenient illusion, but an illusion nonetheless.

To map closets with diverse theoretical ken, at a variety of spatial scales and physical locations, aims at a third implication of this work, which is to treat the closet as a thirdspace: a place that challenges the dualisms in our thinking. It is a place that is both metaphoric and material. It is helpful to understand it as a term that stands for something else, but it is no less useful to see it as a kind of power/knowledge space too. It can be found at any spatial scale (body, city, nation, globe), and can be used (albeit with some caution) in a variety of locations. This is true of 'typically' closeted locations like Japan or the census – but also in ironic locations like New York and London, or one's very own body. As the quote that opened this book shows, it is a term inextricably located in queer culture and politics. Yet it travels – across the globe and throughout other identities. This geography of signification is abetted by the death of the metaphor itself, as more and more people resignify the closet-sign to refer to anything done/known secretly, or in private. It is a form of power/knowledge about which queer theory and praxis is profoundly ambivalent too. As Dan Savage exhorts in Chapter 5, the closet is typically viewed as a bad space, and there is a progressive politics in naming it as an oppression to create solidarity across time and space. But to name it so exclusively – even for the most progressive aims – risks levelling an anti-democratic rendering that makes it hard to see or valorise other aspects of the closet. However much I want to dismantle the closet wherever I find it politically, I must admit that it is a space where amazing things happen. It is also not always mine to name or dismantle. For example, the closet helps impel the heroic struggles that take place within it, the comfort and security many of us have found there, and the liberal democratic principles of privacy and respect and self-determination that underpin so many of our politics in the first place. We must think about the closet as a space to be resisted and dismantled, but simultaneously as a space that is never just the one to be known exclusively through our own privileged geographical imagination.

## Postmodern spaces? Postmodern times, too

Of course this book has its own geography. It was written across a variety of locations, from Canada to New Zealand to the United States. Yet I cannot help but close by considering just how much this book is a product of its *time* as well. The reconceptualisation of the closet offered here reflects and reinforces the poststructural (or postmodern) crisis of representation that has exercised so many scholars at the end of the twentieth century (Sim 1998). To rethink our epistemology of the closet is to ask how we can represent the world if our representations are never pure, always partial, always neglectful or distorting.

148

How do we think beyond the confines of our dualisms: moving from either/or to both/and? Amidst this grand dilemma is a more immediate one that is especially vexing for geographers: what is the 'proper' relationship between reality and our representations of it? The skyrocketing of social theory out of the humanities certainly has brought a demand for consideration of textuality – and empirically oriented disciplines like geography are all the better for it – but where does that leave those of us who are not just interested in texts or their textuality *per se*? On the one hand, the recent ascendance of queer theory in the academy excites and fills me with pride. On the other hand I have traded stories with colleagues about wincing in the back row of humanities conferences when papers dealing with specific places are dismissed as quaintly and naively empiricist. This book certainly does not resolve any of these dilemmas, but it should be read as inevitably situated in them (another trendy spatial metaphor there!). Thus in a recent review of scholarly literature on sexuality and space, Binnie and Valentine (1999: 183) write, 'Geographers could occupy a central place in the rearticulation of queer theory to include a much needed social or material dimension.' I offer this research as a modest part of that vision.

# BIBLIOGRAPHY

Abbas, A. (1996) 'Hyphenation: the spatial dimensions of Hong Kong culture', in M. Steinberg (ed.) *Walter Benjamin and the Demands of History* (Ithaca: Cornell University Press), pp. 214–31.

Abram, S., Murdoch, J. and Marsden, T. (1998) 'Planning by numbers: migration and statistical governance', in P. Boyle and K. Halfacree (eds) *Migration into Rural Areas* (New York: Wiley), pp. 236–51.

Adam, B. (1995) *The Rise of a Gay and Lesbian Movement* (New York: Twayne).

Alcoff, L. (1991) 'The problem of speaking for others', *Cultural Critique*, 20, pp. 5–32.

Aldrich, R. (1993) *The Seduction of the Mediterranean: Writing, Art, and Homosexual Fantasy*. (London: Routledge).

Aldrych, V. C. (1972) 'Pictorial meaning, picture-thinking, and Wittgenstein's theory of aspects', in W. Shibles (ed.) *Essays on Metaphor* (Whitewater, WI: Language Press), pp. 93–110.

Allen, R. E. (ed.) (1990) *Concise Oxford Dictionary of Current English* (Oxford: Clarendon Press).

Altman, D. (1988) 'Legitimization through disaster: AIDS and the gay movement', in E. Fee and D. Fox (eds) *AIDS: The Burdens of History* (Berkeley: University of California Press), pp. 301–15.

—— (1996) 'On global queering', *Australian Humanities Review*, 2, available online at <http://www.lib.latrobe.edu.au/AHR/archive/Issue-July-1996/altman.html

—— (1997) 'Global gaze/global gays', *GLQ: A Journal of Lesbian and Gay Studies*, 3, pp. 417–36.

*The Alyson Almanac* (1993) (Boston: Alyson Press).

Amnesty International (1994) *Breaking the Silence: Human Rights Violations Based on Sexual Orientation* (New York: Amnesty International, 322 Eighth Ave, 10001).

Apter, E. (1996) 'Acting out orientalism', in E. Diamond (ed.) *Performance and Cultural Politics* (London: Routledge), pp. 15–34.

Aristotle [1965], *The Poetics*, trans. T. R. Dorsch, original: 4th century BC (Harmondsworth: Penguin).

—— [1991] *Rhetoric*, trans. G. Kennedy, original 4th century BC (New York: Oxford University Press).

Ark, W. (1997) *Representation Matters: The Effect of 3D Objects and a Spatial Metaphor in a Graphical User Interface* (Yorktown Heights, NY: IBM Research Division), Report no. RJ 10090 (91906).

Asher, R. (1994) *The Encyclopedia of Language and Linguistics* (Oxford: Pergamon).

*Assignment*, NZTV, 25 July 1996.

Atmore, C. (1988) 'Drawing the line: issues of boundary and the homosexual law reform bill campaign in Aotearoa, 1985–86', *Women's Association Conference Papers* (Auckland: Women Students' Association), pp. 162–75.

Austin, J. (1975) *How to Do Things with Words* (New York: Oxford University Press).

Babbie, E. (1997) *The Practice of Social Research*, 8th edition (Belmont, CA: Wadsworth).

Bachelard, G. (1994) *The Poetics of Space* (Boston: Beacon Press).

Backman, G. (1991) *Meaning by Metaphor: An Exploration of Metaphor with a Metaphoric Reading of Two Short Stories by Stephen Crane* (Stockholm: Almqvist and Wiksell International).

Barfield, E. (1960) 'The meaning of the word, "literal"', in C. Knights and B. Cottle (eds) *Metaphor and Symbol* (London: Butterworths), pp. 48–63.

Barnes, T. and Duncan, J. (1992) *Writing Worlds: Discourse, Text, and Metaphor in the Representation of Landscape* (London: Routledge).

Barnhart, R. (1995) *Dictionary of Etymology: The Origins of American English Words* (New York: HarperCollins).

Bawer, B. (ed.) (1996) *Beyond Queer: Challenging Gay Left Orthodoxy* (New York: Free Press).

Bayer, R. (1981) *Homosexuality and American Psychiatry: The Politics of Diagnosis* (Princeton: Princeton University Press).

Beale, P. (ed.) (1989) *A Concise Dictionary of American Slang* (London: Routledge).

Beardsley, M. B. (1972) 'The metaphoric twist', in W. Shibles (ed.) *Essays on Metaphor* (Whitewater, WI: Language Press), pp. 73–93.

Bell, D., Binnie, J., Cream, J. and Valentine, G. (1994) 'All hyped up and no place to go', *Gender, Place, and Culture*, 1, pp. 31–47.

Bell, D. and Valentine, G. (1995a) *Mapping Desire: Geographies of Sexualities* (London: Routledge).

—— (1995b) 'Orientations', in D. Bell and G. Valentine (eds) *Mapping Desire: Geographies of Sexualities* (London: Routledge), pp. 1–27.

—— (1995c) 'The sexed self', in S. Pile and N. Thrift (eds) *Mapping the Subject: Geographies of Cultural Transformation* (London: Routledge), pp. 143–57.

—— (1996) *Consuming Geographies* (London: Routledge).

Benko, G. and Strohmayer, U. (eds) (1997) *Space and Social Theory: Interpreting Modernity and Postmodernity* (Oxford: Blackwell).

Benkov, L. (1994) *Reinventing the Family* (New York: Crown).

Berland, L. and Freeman, E. (1993) 'Queer nationality', in M. Warner (ed.) *Fear of a Queer Planet: Queer Politics and Social Theory* (Minneapolis: University of Minnesota Press), pp. 193–229.

Betsky, A. (1995) *Building Sex: Men, Women, Architecture, and the Construction of Sexuality* (Cambridge, MA: MIT Press).

—— (1997) *Queer Space: Architecture and Same-Space Desire* (New York: Morrow).

Bhabba, H. (1990) *Nation and Narration* (London: Routledge).

Binnie, J. (1997a) 'Coming out of geography: towards a queer epistemology', *Society and Space*, 15, pp. 223–37.

—— (1997b) 'Invisible Europeans: sexual citizenship in the new Europe', *Environment and Planning A*, 29, pp. 237–48.

Binnie, J. and Valentine, G. (1999) 'Geographies of sexuality – a review of progress', *Progress in Human Geography*, 23, pp. 175–87.

Black, M. (1962) *Models and Metaphors* (Ithaca, NY: Cornell University Press).

—— (1979) 'How metaphors work: a reply to Donald Davidson', in S. Sacks (ed.) *On Metaphor* (Chicago: University of Chicago Press).

Bleys, R. (1995) *The Geography of Perversion: Male-to-Male Sexual Behavior Outside the West and the Ethnographic Imagination, 1750–1918* (New York: New York University Press).

Blum, V. and Nast, H. (1996) 'Where's the difference: the heterosexualization of alterity in Henri Lefebvre and Jacques Lacan', *Society and Space*, 14, pp. 559–80.

Blunt, A. (1996) *Travel, Gender, and Imperialism* (New York: Guilford).

Boers, F. (1996) *Spatial Prepositions and Metaphor* (Berlin: Gunter Narr Verlag Tubingen).

Bogue, R. (1989) *Deleuze and Guattari* (London: Routledge).

Bondi, L. (1993) 'Locating identity politics', in M. Keith and S. Pile (eds) *Place and the Politics of Identity* (London: Routledge), pp. 84–101.

Bordo, S. (1992) 'Review essay: postmodern subjects; postmodern bodies', *Feminist Studies*, 18, pp. 159–75.

—— (1993) *Unbearable Weight* (Berkeley: University of California Press).

Bouthillette, A. (1994) 'Gentrification by gay male communities: a case study of Toronto's Cabbagetown', in S. Whittle (ed.) *The Margins of the City: Gay Men's Urban Lives* (Brookfield, VT: Arena Press), pp. 65–84.

—— (1997) 'Queer and gendered housing: a tale of two neighbourhoods in Vancouver', in G. Ingram, A. Bouthillette and Y. Retter (eds) *Queers in Space* (Seattle: Bay Press), pp. 213–32.

Bowie, M. (1979) 'Jacques Lacan', in J. Sturrock (ed.) *Structuralism and Since* (Oxford: Oxford University Press), pp. 116–53.

Boyle, P., Cooke, T., Halfacree, K. and Smith, D. (1999) 'Integrating Great Britain and U.S. census microdata for migration analysis: a study of the effects of family migration on partnered women's employment status', *International Journal of Population Geography* 5, pp. 157–78.

Bright, W. (1992) *The International Encyclopedia of Linguistics* (New York: Oxford University Press).

*Broadsheet* (1993) 'What a difference an act made', Summer, p. 22.

Brooks, P. (1987) 'The idea of a psychoanalytic literary criticism', in S. Rimmon-Kenan (ed.) *Discourse in Psychoanalysis and Literature* (New York: Methuen), pp. 1–18.

Brown, A. (1976) 'Toward a world census', *Population Trends*, 14, pp. 17–19.

Brown, M. (1994) 'The work of city politics: citizenship through employment in the local response to AIDS', *Environment and Planning A*, 26, pp. 873–94.

—— (1995) 'Sex, scale, and the new urban politics: HIV-prevention strategies from

Yaletown, Vancouver', in D. Bell and G. Valentine (eds) *Mapping Desire* (London: Routledge), pp. 245–63.

—— (1996) 'Closet geography', *Society and Space*, 14, pp. 762–70.

—— (1997) *RePlacing Citizenship: AIDS Activism and Radical Democracy* (New York: Guilford).

—— (1999) 'Reconceptualizing public and private in urban regime theory: governance in AIDS politics', *International Journal of Urban and Regional Research*, 23, pp. 70–87.

Browning, F. (1996) *A Queer Geography: Journeys toward a Sexual Self* (New York: Crown Publishers).

Burg, B. (1983) *Sodomy and the Pirate Tradition: English Sea Rovers in the Seventeenth Century Caribbean* (New York: New York University Press).

Busch, A. (1999) *Geography of Home* (Princeton: Princeton Architectural Press).

Butler, J. (1990) *Gender Trouble* (London: Routledge).

—— (1991) 'Imitation and gender insubordination', in D. Fuss (ed.) *Inside/Out* (London: Routledge), pp. 13–31.

—— (1993a) 'Critically queer', *GLQ: A Journal of Lesbian and Gay Studies*, 1, pp. 17–32.

—— (1993b) *Bodies That Matter* (London: Routledge).

—— (1996) 'Performativity's social magic', in T. Schatzki and W. Natter (eds) *The Social and Political Body* (New York: Guilford).

—— (1997a) 'Sovereign performatives in the contemporary scene of the utterance', *Critical Inquiry*, 23, pp. 350–77.

—— (1997b) *Excitable Speech: A Politics of the Performative* (London: Routledge).

—— (1997c) *The Psychic Life of Power* (Stamford: Stamford University Press).

Calfia, P. (1995) *Public Sex: The Culture of Radical Sex* (Pittsburgh: Cleis Press).

Calhoun, C. (1995) 'The gender closet: lesbian disappearance under the sign "women"', *Feminist Studies*, 21, pp. 7–34.

Cant, B. (ed.) (1997) *Invented Identities: Lesbians and Gays Talk about Migration* (London: Cassell).

Canterbury Tourism Council (1995) 'Christchurch and Canterbury Itinerary Planner's Guide' (CTC, PO Box 2600, Christchurch, NZ).

Caplan, C. (1998) 'On location', in S. Aiken, A. Brigham, S. Marston and P. Waterstone (eds) *Making Worlds: Gender, Metaphor, and Materiality* (Tucson: University of Arizona Press), pp. 60–5.

Carter, V. (1995) 'Abseil makes the heart grow fonder: lesbian and gay campaigning tactics and Section 28', in L. Duggan and N. Hunter (eds) *Sex Wars: Sexual Dissent and Political Culture* (London: Routledge), pp. 217–26.

Case, S. E. (1996) *The Domain-Matrix: Performing Lesbian at the End of Print Culture* (Bloomington: University of Indiana Press).

Case, S. E., Brett, P. and Foster, S. L. (1995) *Cruising the Performative: Interventions into the Representation of Ethnicity, Nationality, and Sexuality* (Bloomington: Indiana University Press).

Cassier, E. (1953) *Language and Myth* (New York: Dover).

Castells, M. (1983) *The City and the Grassroots* (Berkeley: University of California Press).

Chapman, R. (1986) *Dictionary of American Slang* (New York: Harper and Row).

Chauncey, G. (1994) *Gay New York* (New York: Basic Books).

Chiappe, D. (1998) 'Similarity, relevance, and the comparison process', *Metaphor and Symbolic Activity*, 13, pp. 17–30.

Chouinard, V. and Grant, A. (1996) 'On being not even anywhere near 'the project': ways of putting ourselves in the picture', in N. Duncan (ed.) *BodySpace* (London: Routledge), pp. 170–96.

Christchurch City Council (1996) *1996/7 Annual Plan* (PO Box 237, Christchurch, NZ).

*The Christchurch Mail* (Christchurch, New Zealand) (1996) 'Existing "Massage" sign approved', 3 August, p.1.

*The Christchurch Star* (Christchurch, New Zealand) (1996) 'Sleazy label rejected', 24 June, p. 1.

—— (1996) 'Sleazy bylaw plan falters', 9 August, p. 1.

Clark, W. A. V. and Hoskins, P. (1986) *Statistical Methods for Geographers* (New York: John Wiley and Sons).

Clifford, J. (1992) 'Traveling cultures', in L. Grossberg, C. Nelson and P. Treichler (eds) *Cultural Studies* (London: Routledge), pp. 96–116.

Cloke, P., Philo, C. and Sadler, D. (1993) *Approaching Human Geography* (New York: Guilford).

Cohen, E. (1996) 'Posing the question: Wilde, wit, and the ways of man', in E. Diamond (ed.) *Performance and Cultural Politics* (London: Routledge), pp. 35–47.

Collins, J. and Mayblin, B. (1996) *Derrida for Beginners* (Cambridge: Icon).

Cooper, D. (1986) *Metaphor*, 5, Aristotelian Society Series (Oxford: Basil Blackwell).

—— (1994) *Power in Struggle: Feminism, Sexuality, and the State* (Milton Keynes: Open University Press).

Crang, E. (1998) *The Routledge Encyclopedia of Philosophy* (London: Routledge).

Crang, M., Crang, P. and May, J. (1999) *Virtual Geographies: Bodies, Spaces, and Relations* (London: Routledge).

Cream, J. (1995a) 'Re-solving riddles: the sexed body', in D. Bell and G. Valentine (eds) *Mapping Desire* (London: Routledge), pp. 31–40.

—— (1995b) 'Women on trial: a private pillory?' in S. Pile and N. Thrift (eds) *Mapping the Subject* (London: Routledge), pp. 158–69.

Cresswell, T. (1996) *In Place/Out of Place: Geography, Ideology, and Transgression* (Minneapolis: University of Minnesota Press).

—— (1997) 'Weeds, plagues, and bodily secretions: a geographical interpretation of metaphors of displacement', *Annals of the Association of American Geographers*, 87, pp. 330–45.

Crimp, D. (1994) 'Right on, girlfriend!' in M. Warner (ed.) *Fear of a Queer Planet: Queer Politics and Social Theory* (Minneapolis: University of Minnesota Press), pp. 300–20.

Cruikshank, M. (1992) *The Gay and Lesbian Liberation Movement* (New York: Routledge).

Curry, M. (1996) *The Work in the World: Geographic Practice and the Written Word* (Minneapolis: University of Minnesota Press).

Dale, A. and Marsh, C. (1993) *The 1991 Census User's Guide* (London: HMSO).

Damle, S. (1997) 'Three dimensional spatial metaphor for interaction in cyberspace', unpublished MS thesis, Washington State University.

Darier, E. (1996) 'The politics and power effects of garbage recycling in Halifax, Canada', *Local Environment*, 1, pp. 63–86.

Delaney, D. and Leitner, H. (1997) 'The political construction of scale', *Political Geography*, 16, pp. 93–7.

Deleuze, G. (1962) *Nietzsche and Philosophy*, trans. H. Tomlinson (New York: Columbia University Press).

Deleuze, G. and Guattari, F. (1983 [1972]) *Anti-Oedipus: Capitalism and Schizophrenia*, trans. R. Hurley, M. Seem and H. Rane (Minneapolis: University of Minnesota Press).

Derrida, J. (1972) *Dissemination* (London: Athlone Press).

—— (1978) 'The *retrait* of metaphor', *Enclitic*, 11, pp. 5–33.

—— (1982) *Margins of Philosophy*, trans. A. Bass (Chicago: University of Chicago Press).

Diamond, E. (1996) 'Introduction', in E. Diamond (ed.) *Performance and Cultural Politics* (London: Routledge), pp. 1–12.

Dickerson, G. (1996) 'Festivities and jubilations on the graves of the dead: sanctifying sullied space', in E. Diamond (ed.) *Performance and Cultural Politics* (London: Routledge), pp. 108–27.

Dickey, J. (1968) *Metaphor as Pure Adventure* (Washington: Library of Congress).

Dirven, R. (1983) 'Metaphors of spatial relations', *Trierer Studien zur Literatur*, 7, pp. 63–91.

*The Dominion* (Wellington, New Zealand) (1996) 'Use prostitution to draw tourists', 4 November, p. 1.

Dreeuws, D. (1998) 'Taking up space: transgender lived experience and metaphors of space and place', unpublished MA thesis, Claremont University.

Duberman, M. (1994) *Stonewall* (New York: Plume).

Duggan, L. (1995) 'Queering the state', in L. Duggan and N. Hunter (eds) *Sex Wars: Sexual Dissent and Political Culture* (London: Routledge), pp. 179–93.

Duncan, J. (1990) *The City as Text* (Cambridge: Cambridge University Press).

Duncan, J. and Duncan, N. (1992) 'Ideology and bliss: Roland Barthes and the secret histories of landscape', in T. Barnes and J. Duncan (eds) *Writing Worlds* (London: Routledge), pp. 18–37.

Duncan, J. and Gregory, D. (1999) *Writes of Passage: Reading Travel Writing* (London: Routledge).

Duncan, N. (1996) 'Renegotiating gender and sexuality in public and private spaces', in N. Duncan (ed.) *BodySpace* (London: Routledge), pp. 127–45.

Eagleton, T. (1983) *Literary Theory: An Introduction* (Minneapolis: University of Minnesota Press).

—— (1990) *The Significance of Theory* (Oxford: Blackwell).

Ebensten, H. (1993) *Volleyball with the Cuna Indians and Other Gay Travel Adventures* (New York: Penguin).

Edelman, L. (1994) *Homographesis* (London: Routledge).

Eden, P. (1997) 'Lust, trust, and money: an economic sociology of the sex industry in Christchurch, New Zealand', unpublished MA thesis, University of Canterbury, NZ.

Edwards, T. (1998) 'What the cutters feel', *Time*, vol. 152, no. 19, 9 November 1998, p. 93.

Elder, G. (1998) 'The South African body politic: space, race and heterosexuality', in H. Nast and S. Pile (eds) *Places through the Body* (London: Routledge), pp. 153–64.

Elmer, G. (1996) 'Of Deleuze, diagrams, data and space: or mapping techniques of Foucaultian governmentality, unpublished MA thesis, University of Massachusetts-Amherst.

Embler, W. (1966) *Metaphor and Meaning* (DeLand, FL: Everett/Edwards).

Fairclough, N. (1995) *Critical Discourse Analysis* (New York: Longman).

Ferguson, G. (1994) *Building the New Zealand Dream* (Palmerston North, NZ: The Dunmore Press).

Fitzgerald, T. (1993) *Metaphors of Identity: A Culture–Communication Dialog* (Albany: SUNY Press).

Fleischman, S. (1991) 'Discourse as space/discourse as time: reflections on the meta-language of spoken and written discourse', *Journal of Pragmatics*, 16, pp. 291–306.

*Forbidden Passages: Writings Banned in Canada* (1995) (San Francisco: Cleis Press).

Forsyth, A. (1997) 'Out in the valley', *International Journal of Urban and Regional Research*, 21, pp. 38–62.

Foucault, M. (1980) in C. Gordon (ed.) *Power/Knowledge* (New York: Pantheon).

—— (1991) 'Governmentality', in G. Burchell, C. Gordon and P. Miller (eds) *The Foucault Effect: Studies in Governmentality* (Chicago: University of Chicago Press).

Frantzen, A. (1998) *Before the Closet* (Chicago: University of Chicago Press).

Fuery, P. (1995) *Theories of Desire* (Melbourne: Melbourne University Press).

Fuller, J. and Blackley, S. (1995) *Restricted Entry: Censorship on Trial* (Vancouver: Press Gang Publishers).

Fuss, D. (ed.) (1991) *Inside/Out: Lesbian Theories, Gay Theories* (New York: Routledge).

Gallop, J. (1985) *Reading Lacan* (Ithaca, NY: Cornell University Press).

*The Gay Almanac* (1996) (New York: Berkeley Press).

Gearing, N. (1997) *Emerging Tribe: Gay Culture in New Zealand in the 1990s* (Auckland: Penguin).

Gibson-Graham, J. K. (1997) *The End of Capitalism (As We Knew It)* (Oxford: Blackwell).

Giddens, A. (1984) *The Constitution of Society* (Berkeley: University of California Press).

—— (1987) *The Nation-State and Violence* (Berkeley: University of California Press).

Gmunder, B. (ed.) (1998) *Spartacus Guide for Gay Men '98/99* (Berlin: Bruno Gmunder).

Gonzales, D. (1997) 'What's the problem with "hispanic"? Just ask a "Latino"', in S. Frenkel (ed.) *Introduction to Geography: A Reader* (New York: American Heritage Custom Publishing), pp. 125–7.

Gordon, C. (1991) 'Governmental rationality: an introduction', in G. Burchell, C. Gordon and P. Miller (eds) *The Foucault Effect: Studies in Governmentality* (Chicago: University of Chicago Press).

Gordon, P. (1990) 'The enigma of Aristotelian metaphor: a deconstructive analysis', *Metaphor and Symbolic Activity* 5, pp. 83–90.

Graham, P. (1995) 'Girls' camp?: The politics of parody', in T. Wilton (ed.) *Immortal, Invisible: Lesbians and the Moving Image* (London: Routledge), pp. 163–81.

Gregory, D. (1994) *Geographical Imaginations* (Oxford: Blackwell).

—— (1995) 'Between the book and the lamp: imaginative geographies of Egypt 1849–1950', *Transactions of the Institute of British Geographers*, 20, pp. 29–57.

—— (1997) 'Lacan and geography: the production of space revisited', in G. Benko and U. Strohmeyer (eds) *Space and Social Theory* (Oxford: Blackwell), pp. 203–34.

Gregory, D. and Urry, J. (1985) *Social Relations and Spatial Structures* (New York: St. Martin's Press).

Gregson, N. and Rose, G. (2000) 'Taking Butler elsewhere: performativities, spatialities, and subjectivities', *Society and Space*, 18 (in press).

Grice, H. (1989) *Studies in the Way of Words* (Cambridge, MA: Harvard University Press).

Gross, L. (1993) *Contested Closets* (Minneapolis: University of Minnesota Press).

Grosz, E. (1990) *Jacques Lacan: A Feminist Introduction* (London: Routledge).

Hacking, I. (1991) 'How should we do the history of statistics?', in G. Burchell, C. Gordon and P. Miller (eds) *The Foucault Effect: Studies in Governmentality* (Chicago: University of Chicago Press).

Halberstram, J. (1994) 'F2M: the making of female masculinity', in L. Doan (ed.) *The Lesbian Postmodern* (New York: Columbia University Press), pp. 210–28.

Hall Carpenter Archives (ed.) (1989) *Walking after Midnight: Gay Men's Life Stories* (London: Routledge).

Halperin, D. (1995) *Saint Foucault* (New York: Oxford University Press).

Handelsman, R. (1974) 'Spatial metaphors in Arnold Bennett', unpublished Ph.D. dissertation, University of Illinois, Urbana Champaign, IL.

Hane, A. (1995) 'Swimming upstream and succeeding: academic mothers' use of spatial metaphors in descriptions of self', unpublished Ph.D. dissertation, University of Kansas, Lawrence, KS.

Hanenberg, J. (1982) 'The place of dreams: a qualitative study of the metaphors of space in dreams', unpublished Ed.D. dissertation, Boston University, Boston, MA.

Hanson, S. and Pratt, G. (1997) *Gender, Work, and Space* (London: Routledge).

Harlan, R. (1991) *Superstructuralism* (London: Routledge).

Harris, K. (1996) 'My life in the sex industry', *The Press* (Christchurch, New Zealand), 10 August, p. W-1.

Harvey, D. (1985) *The Urbanization of Capital* (Baltimore: Johns Hopkins University Press).

—— (1989) *The Condition of Postmodernity* (Oxford: Blackwell).

—— (1996) 'The geography of capitalist accumulation', in J. Agnew, D. Livingstone and A. Rogers (eds) *Human Geography: An Essential Anthology* (Oxford: Blackwell).

Hawkes, T. (1972) *Metaphor* (London: Methuen).

Henderson, M. (ed.) (1995) *Borders, Boundaries, and Frames: Cultural Criticism and Cultural Studies* (London: Routledge).

Herod, A. (1997) 'Labor's spatial praxis and the geography of contract bargaining in the US east coast longshore industry, 1953–89', *Political Geography*, 16, pp. 145–69.

Hetherington, K (1998) *Expressions of Identity: Space, Performance, Politics* (London: Sage).

Higonnet, M. (1994) 'Mapping the text: critical metaphors', in M. Higonnet and J. Templeton (eds) *Reconfigured Spheres: Feminist Explorations of Literary Space* (Amherst: University of Massachusetts Press), pp. 194–212.

Hilson, M. (1996) 'Homophobia and postcolonialsim', in Emory University Postcolonial

Studies (Atlanta, GA: Emory University), available online at http://www.emory.edu/ENGLISH/Bahri/homophobia.HTML.

Hocquenghem, G. (1993 [1972]) *Homosexual Desire*, trans. D. Dangoor (Durham: Duke University Press).

Hogan, S. and Hudson, L. (1998) *Completely Queer: The Gay and Lesbian Encyclopedia* (New York: Henry Holt).

hooks, b. (1984) *Feminist Theory from Margin to Center* (Boston: South End Press).

Hunter, N. (1995) 'Life after Hardwick', in L. Duggan and N. Hunter (eds) *Sex Wars: Sexual Dissent and Political Culture* (London: Routledge), pp. 85–100.

IBM Research Division (1997) 'Representation matters: the effect of 3D objects and a spatial metaphor in a graphical user interface', RJ 10090, 6 November (Yorktown Heights: IBM Research Division).

Imbriglio, C. (1995) '"Our days put on such reticence": the rhetoric of the closet in John Ashbery's *Some Trees*', *Contemporary Literature*, 36, pp. 249–88.

Indurkhya, B. (1994) 'The thesis that all knowledge is metaphorical and meanings of metaphor', *Metaphor and Symbolic Activity*, 9, pp. 61–73.

Ingram, G., Bouthillette, A. and Retter, Y. (1997) *Queers in Space* (Seattle: Bay Press).

Jacobs, J. M. (1996) *Edge of Empire: Postcolonialism and the City* (London and New York: Routledge).

Jagose, A. (1996) *Queer Theory* (Dunedin: University of Otago Press).

Jakobson, R. (1990) 'Two aspects of language and two types of aphasic disturbances', in L. Waugh and M. Monville-Burston (eds) *On Language* (Cambridge, MA: Harvard University Press), pp. 115–33.

James, D. (1960) 'Metaphor and symbol', in L. Knights and B. Cottle (eds) *Metaphor and Symbol* (London: Butterworth), pp. 95–103.

Jarosz, L. (1992) 'Constructing the dark continent: metaphor as geographic representation of Africa', *Geografiska Annaler*, 74, pp. 105–15.

Jay, K. and Young, A. (1972) *Out of the Closets: Voices of Gay Liberation* (New York: Douglas).

Jay, M. (1993) *Downcast Eyes: The Denigration of Vision in Twentieth-Century French Thought* (Berkeley: University of California Press).

Jefferey, S. (1994) 'The queer disappearance of lesbians: sexuality in the academy', *Women's Studies International Forum*, 17, pp. 459–72.

Johnson, F. (1996) *Geography of the Heart: A Memoir* (New York: Washington Square Press).

Johnson, M. (1995) 'Why metaphor matters to philosophy', *Metaphor and Symbolic Activity*, 10, pp. 157–62.

Johnston, L. (1996) 'Flexing femininity: female body-builders refiguring "the body"', *Gender, Place, and Culture*, 3, pp. 327–40.

Johnston, R. (1997) *Geography and Geographers: Anglo-American Human Geography Since 1945*, 5th edition (London: Arnold).

Jones, A. (1997) 'Teaching post-structuralist feminist theory in education: student resistances', *Gender and Education*, 9, pp. 261–9.

Jones, R. S. (1982) *Physics as Metaphor* (Minneapolis: University of Minnesota Press).

Kaplan, C. (1998) 'On location', in S. Aiken, A. Brigham, S. Marston and P. Waterstone (eds) *Making Worlds: Gender, Metaphor, and Materiality* (Tucson: University of Arizona Press), pp. 60–5.

Kearns, G. (1992) 'Historical geography', *Progress in Human Geography*, 16, pp. 406–13.

Keenan, D. (1996) 'Police keep records on sex cruisers', *The Press* (Christchurch, New Zealand), 16 February, p. A-1.

Kelsey, J. (1995) *The New Zealand Experiment* (Auckland: Auckland University Press).

Kidron, M. and Segal, R. (1984) *The New State of the World Atlas* (New York: Penguin).

Kim, H. and Hirtle, S. (1995) 'Spatial metaphors and disorientation in hypertext browsing', *Behaviour and Information Technology*, 14, pp. 239–50.

King, J. (1991) 'Models as metaphors', *Metaphor and Symbolic Activity*, 6, pp. 103–18.

Kinsman, G. (1987) *The Regulation of Desire* (Montreal: Black Rose Books).

Kirby, A. (1995) 'Straight talk on the pomohomo question', *Gender, Place, and Culture*, 2, pp. 89–95.

Kirby, K. (1996) *Indifferent Boundaries: Spatial Concepts of Human Subjectivity* (New York: Guilford).

Kirby, S. and Hay, I. (1997) '(Hetero)sexing space: gay men and "straight" space in Adelaide, Australia', *Professional Geographer*, 49, pp. 295–305.

Knopp, L. (1990) 'Some theoretical implications of gay involvement in an urban land market', *Political Geography Quarterly*, 9, pp. 337–52.

—— (1992) 'Sexuality and the spatial dynamics of capitalism', *Society and Space*, 10, pp. 651–69.

—— (1995a) 'Sexuality and urban space: a framework for analysis', in D. Bell and G. Valentine (eds) *Mapping Desire* (London: Routledge), pp. 149–64.

—— (1995b) 'If you're going to get all hyped up, you'd better go somewhere', *Gender, Place, and Culture*, 2, pp. 85–9.

Knox, P. and Marston, S. (1998) *Human Geography: Places and Regions in a Global Context* (Upper Saddle River, NJ: Prentice Hall).

Koeksen, H. (1993) 'Christchurch city needs an urban vision', *Planning Quarterly*, 112, pp. 22–4.

LaBruce, B. and Belverio, G. (1998) 'A case for the closet', in M. Simpson (ed.) *Anti-Gay* (London: Freedom Editions), pp. 140–63.

Lacan, J. (1977) *Ecrits: A Selection*, trans. A. Sheridan (New York: Norton).

—— (1994 [1973]) *The Four Fundamental Concepts of Psychoanalysis*, trans. A. Sheridan (Harmondsworth: Penguin).

Lacapra, D. (1980) 'Who rules metaphor?', *diacritics*, 10, pp. 15–28.

Lakoff, G. (1987) 'The death of a dead metaphor', *Metaphor and Symbolic Activity*, 2, pp. 143–7.

Lakoff, G. and Johnson, M. (1980) *Metaphors We Live By* (Chicago: University of Chicago Press).

Lane, C. (1999) 'Response to Denis Altman', *Australian Humanities Review* <http://www.lib.latrobe.edu.au/AHR/emuse/Globalqueering/lane.html

Langston, W. (1994) 'The use of spatial metaphors when thinking about nonspatial domains', unpublished Ph.D. dissertation, University of Wisconsin, Madison, WI.

Lapovsky-Kennedy, E. and Davis, M. (1993) *Boots of Leather, Slippers of Gold: The History of a Lesbian Community* (New York: Routledge).

Laumann, E., Gagnon, J., Michael, R. and Michael, S. (1994) *The Social Organization of Sexuality: Sexual Practices in the United States* (Chicago: University of Chicago Press).

Lawn, K. (1995) 'Pumping life into the inner city', *Planning Quarterly*, 114, pp. 26–8.

Leader, D. and Groves, J. (1995) *Lacan for Beginners* (London: Icon).

Leddy, T. (1995) 'Metaphor and metaphysics', *Metaphor and Symbolic Activity*, 10, pp. 205–22.

Leeman, R. (1995) 'Spatial metaphors in African-American discourse', *Southern Communication Journal*, 60, pp. 165–71.

Lefebvre, H. (1991) *The Production of Space*, trans. D. Nicholson-Smith (Oxford, UK and Cambridge, MA: Blackwell).

—— (1996) *Writings on Cities*, trans. E. Kofman and E. Lebas (Oxford, UK and Cambridge, MA: Blackwell).

Levenson, M. (1998) 'Speaking to power: the performances of Judith Butler', *Lingua Franca*, 8, pp. 60–9.

Lewes, K. (1988) *The Psychoanalytic Theory of Male Homosexuality* (New York: Meridian).

Ley, D., Hiebert, D. and Pratt, G. (1992) 'Time to grow up? From urban village to world city 1966–1991', in G. Wynn and T. Oke (eds) *Vancouver and Its Region* (Vancouver: University of British Columbia Press), pp. 234–66.

Luke, P. (1993) 'Gay law passes easily after it becomes a health issue', *The Press* (Christchurch, New Zealand), p. 22.

Luke, T. (1994) 'Placing power/siting space: the politics of local and global in the New World Order', *Society and Space*, 12, pp. 613–28.

MacAdam, A. J. (1980) 'Origins and narratives', *Modern Language Notes*, 95, pp. 424–35.

McCarthy, J. (1994) 'The closet and the ethics of outing', *Journal of Homosexuality*, 27, pp. 17–45.

McDowell, L. (1995) 'Body work: heterosexual gender performances in city workplaces', in D. Bell and G. Valentine (eds) *Mapping Desire* (London: Routledge), pp. 75–98.

McKaym, D. (1985) 'Personification and spatial metaphors in literature for children', *Metaphor*, 1, pp. 87–107.

MacKay, D. (1986) 'Prototypicality among metaphors: on the relative frequency of personification and spatial metaphors in literature written for children versus adults', *Metaphor and Symbolic Activity*, 1, pp. 87–107.

McNamara, R. P. (1994) *Times Square Hustler: Male Prostitution in New York City* (Westport, CT: Prager Press).

Malmkjaer, K. (1991) *The Linguistics Encyclopedia* (London: Routledge).

Mann, K. (1982) 'Self, shell, and world: George Eliot's language of space', *Genre*, 15, pp. 447–76.

Marcus, E. (ed.) (1992) *Making History* (New York: HarperCollins).

—— (1993) *Is It a Choice?* (New York: HarperCollins).

Martens, L. (1982) 'Irreversible processes, proliferating middles, and invisible barriers: spatial metaphors in Freud, Schnitzler, Musil, and Kafka', in E. Haymes (ed.) Focus on

Vienna 1900: Change and Continuity in Literature, Music, Art, and Intellectual History, *Houston German Studies*, 4, pp. 46–57.

Massey, D. (1984) *Spatial Divisions of Labor* (London: Methuen).

—— (1993) *Space, Place, and Gender* (Minneapolis: University of Minnesota Press).

Mathur, T. (1996) 'Metaphor of space: Charlotte Keatley's play *My Mother Said I Never Should*,' in J. Jain (ed.) *Women's Writing: Text and Context* (New Delhi: Rawat Publications), pp. 287–94.

Meisel, M. (1979) 'Waverly, Freud, and topographical metaphor', *University of Toronto Quarterly*, 48, pp. 226–44.

Mercken-Spaas, G. (1977) 'The metaphor of space in Constant's "Adolphe"', *Nineteenth Century French Studies*, 3–4, pp. 186–95.

Merrifield, A. (1997) 'Between process and individuation: translating metaphors and narratives of urban space', *Antipode*, 29, pp. 417–36.

Michael, R., Gagnon, J., Laumann, E. and Kolata, G. (1994) *Sex in America: A Definitive Study* (Boston: Little, Brown).

Miller, N. (1989) *In Search of Gay America: Women and Men in a Time of Change* (New York: Harper and Row).

—— (1992) *Out in The World: Gay and Lesbian Life from Buenos Aires to Bangkok* (New York: Vintage).

Miller, P. and Rose, N. (1990) 'Governing economic life', *Economy and Society*, 19, pp. 1–31.

Mitchell, T. (1990) 'Everyday metaphors of power', *Theory and Society*, 19, pp. 545–75.

Mohr, R. (1988) *Gays/Justice* (New York: Columbia University Press).

—— (1992) *Gay Ideas: Outing and Other Controversies* (Boston: Beacon Press).

Moon, M. (1993) 'New Introduction' to G. Hocquenghem, *Homosexual Desire*, originally published 1972 (Chapel Hill: Duke University Press), pp. 9–21.

Moon, M., Sedgwick, E., Giannil, B. and Weir, S. (1994) 'Queers in (single-family) space', *Assemblage: A Critical Journal of Architecture and Design Culture*, 24, pp. 30–7.

Morton, D. (ed.) (1996) *The Material Queer: A Lesbigay Cultural Studies Reader* (Boulder, CO: Westview Press).

Murdoch, J. (1997a) 'The shifting territory of government: some insights from the Rural White Paper', *Area*, 29, pp. 109–18.

—— (1997b) 'Governmentality and territoriality: the statistical manufacture of Britain's "national farm"', *Political Geography*, 16, pp. 307–24.

Murray, S. (1984) 'Social theory, homosexual realities', Gai Saber Monograph 3 (New York: Gay Academic Union).

Murry, J. M. (1972) 'Metaphor', in W. Shibles (ed.) *Essays on Metaphor* (Whitewater, WI: Language Press), pp. 27–40.

Nardi, P., Sanders, D. and Marmor, J. (eds) (1994) *Growing up before Stonewall: Life Stories of Some Gay Men* (London: Routledge).

Nast, H. and Pile, S. (eds) *Places through the Body* (London: Routledge).

National Lesbian and Gay Survey (ed.) (1993) *Proust, Cole Porter, Michelangelo, Marc Almond and Me: Writings by Gay Men on Their Lives and Lifestyles* (London: Routledge).

Natter, W. and Jones, J. P. (1993) 'Signposts towards a poststructuralist geography', in

J. P. Jones, W. Natter and T. Schatzki (eds) *Postmodern Contentions: Epochs, Politics, Space* (New York: Guilford), pp. 165–203.

Nelson, L. (1999) 'Bodies (and spaces) do matter: the limits of performativity', *Gender, Place, and Culture*, 6, pp. 331–54.

*New Zealand Statutes* (1961) 'Crimes against Public Welfare', Section 146, p. 67.

—— (1961) 'Crimes Act', Sections 147–9, pp. 67–8.

—— (1978) 'Massage Parlor Act', Section 13, p. 82.

—— (1986) 'Homosexual Law Reform', Section 6, p. 463.

Nietzsche, F. (1972) 'On the truth and falsity in their extramural sense', in W. Shibles (ed.) *Essays on Metaphor* (Whitewater, WI: Language Press), pp. 1–14.

Noppen, J. P. (1974) 'Spatial metaphors in contemporary British religious prose', *Revue des Langues Vivantes*, 40, pp. 7–24.

Office of Population Censuses and Surveys (1993) 'In brief', *Population Trends*, 72, pp. 1–9.

O'Malley, P., Weir, L. and Shearing, C. (1997) 'Governmentality, criticism, politics', *Economy and Society*, 26, pp. 501–17.

Ortnoy, A. (ed.) (1979) *Metaphor and Thought* (Cambridge: Cambridge University Press).

Osborn, T. (1996) *Coming Home: A Roadmap to Gay and Lesbian Empowerment* (New York: St. Martin's Press).

—— (1997) 'Of health and statecraft', in A. Peterson and R. Bunton (eds) *Foucault, Health, and Medicine* (London: Routledge), pp. 173–88.

Owen, R. (1996) 'The population census of 1917 and its relationship to Egypt's three 19th century statistical regimes', *Journal of Historical Sociology*, 9, pp. 457–72.

Painter, J. (1996) *Politics, Geography, and Political Geography* (London: Arnold).

Pallister, D. and Gibbs, G. (1998) 'Outing tests press codes', *The Guardian*, 10 November, p. 3.

Parker, A., Russo, M., Sommer, D. and Yaeger, P. (1992) *Nationalisms and Sexualities* (New York: Routledge).

Parker, A. and Sedgwick, E. (1995) *Performativity and Performance* (London: Routledge).

Parkinson, P. (1989) 'Sexual law reform: the New Zealand Experience from Wolfenden to the Crimes Bill 1989', *Sites*, 19, pp. 7–13.

Patton, C. (1995) 'Performativity and spatial distinction', in A. Parker and E. Sedgwick (eds) *Performativity and Performance* (London: Routledge), pp. 173–96.

Peet, R. (1991) *Global Capitalism: Theories of Societal Development* (New York and London: Routledge).

Person, L. (1980) 'Playing house: Jane Austen's fabulous space', *Philological Quarterly*, 59, pp. 62–75.

Petersen, A. (1997) 'Risk, governance, and the new public health', in A. Peterson and R. Bunton (eds) *Foucault, Health, and Medicine* (London: Routledge), pp. 189–206.

Philbrick, A. (1979) 'Space as metaphor and theater: the individual's search for place in selected works of Colette', unpublished Ph.D. dissertation, Brown University, Providence, RI.

Phillips, R. (1999) 'Writing travel and mapping sexuality: Richard Burton's Sotatic Zone', in J. Duncan and D. Gregory (eds) *Writes of Passage: Reading Travel Writing* (London: Routledge), pp. 70–91.

Pile, S. (1996) *The Body and the City* (London: Routledge).

Pile, S. and Thrift, N. (1995) *Mapping the Subject: Geographies of Cultural Transformation* (London: Routledge).

Plummer, K. (1995) *Telling Sexual Stories: Power, Change, and Social Worlds* (London: Routledge).

Pollack, D. (1992) 'Forced out of the closet: sexual orientation and the legal dilemma of "outing"', *University of Miami Law Review*, 46, pp. 711–50.

Porter, K. and Weeks, J. (eds) (1991) *Between the Acts: Lives of Homosexual Men 1885–1967* (London: Routledge).

Pratt, G. (1992) 'Spatial metaphors and speaking positions', *Society and Space*, 10, pp. 241–4.

—— (1998) 'Geographic metaphors in feminist theory', in S. Aiken, A. Brigham, S. Marston and P. Waterstone (eds) *Making Worlds: Gender, Metaphor, and Materiality* (Tucson: University of Arizona Press), pp. 13–30.

—— (2000) 'Performativity', in R. Johnston (ed.) *The Dictionary of Human Geography*, 4th edition (Oxford: Blackwell), in press.

Preston, J. (ed.) (1991) *Hometowns: Gay Men Write about Where They Belong* (New York: Dutton).

—— (ed.) (1992) *A Member of the Family: Gay Men Write about Their Families* (New York: Dutton).

—— (ed.) (1995) *Friends and Lovers: Gay Men Write about the Families They Create* (New York: Plume).

Price-Chalita, P. (1994) 'Spatial metaphor and the politics of empowerment: mapping a place for feminism and postmodernism in geography?', *Antipode*, 26, pp. 236–54.

Probyn, E. (1995) 'Lesbians in space: sex and the structure of missing', *Gender, Place, and Culture*, 2, pp. 77–83.

Propp, V. (1968) *Morphology of the Folktale* (Austin: University of Texas Press).

Radden, G. (1985) 'Spatial metaphors underlying prepositions of causality', in R. Driver and W. Paprotte (eds) *The Ubiquity of Metaphor: Metaphor in Language and Thought* (Amsterdam: John Benjamins) (C.I.L.T. vol. 29), pp. 177–207.

Radman, Z. (1997) 'Difficulties with diagnosing the death of a metaphor', *Metaphor and Symbolic Activity*, 12, pp. 149–57.

Ranciere, J. (1994) 'Discovering new worlds: politics of travel and metaphors of space', in G. Robertson *et al.* (eds) *Travellers Tales: Narratives of Home and Displacement* (London: Routledge), pp. 29–37.

Renton, N. (1990) *Metaphors: An Annotated Dictionary* (Melbourne: Schwartz and Wilkinson).

Richards, I. A. (1965 [1936]) *The Philosophy of Rhetoric* (Oxford: Oxford University Press).

Richards, S. (1995) 'Writing the absent potential: drama, performance, and the canon of African-American Literature', in A. Parker and E. Sedgwick (eds) *Performativity and Performance* (London: Routledge), pp. 84–8.

Ricoeur, P. (1978) *The Rule of Metaphor*, trans. R. Czerny (London: Routledge and Kegan Paul).

Rist, D. (1993) *Heartlands* (New York: Plume).

Roach, J. (1995) 'Culture and performance in the cirum-Atlantic world', in A. Parker and E. Sedgwick (eds) *Performativity and Performance* (London: Routledge), pp. 45–63.

Rodi, R. (1994) *Closet Case* (New York: Plume).

Roediger, H. (1980) 'Memory metaphors in cognitive psychology', *Memory and Cognition*, 8, pp. 231–46.

Rose, G. (1993) *Feminism and Geography: The Limits of Geographical Knowledge* (Minneapolis, MN: University of Minnesota Press).

—— (1996) 'As if the mirror has bled: masculine dwelling, masculine theory, and feminist masquerade', in N. Duncan (ed.) *BodySpace* (London: Routledge), pp. 56–74.

Rozik, E. (1994) 'Stage metaphor: with or without. On Rina Yerushalmi's production of Ionesco's *The Chairs*', *Theatre Research International*, 19, pp. 148–55.

Ryle, G. (1955) *The Concept of Mind* (London: Hutchinson).

Sack, K. (1998) 'Georgia's high court voids sodomy law', *New York Times*, 24 November, p. 1.

Said, E. (1983) *The Word, the Text, and the Critic* (Cambridge, MA: Harvard University Press).

Sample, M. (1991) 'Landscape and spatial metaphor in Bessie Head's *The Collector of Treasures*', *Studies in Short Fiction*, 28, pp. 311–20.

Sarup, M. (1989) *An Introductory Guide to Postmodernism and Poststructuralism* (Sydney: Harvester Wheatsheaf Press).

Saunders, P. (1980) 'Towards a non-spatial urban sociology', Working Paper 21, Urban and Regional Studies Department, University of Sussex, Falmer, Brighton, BN1 9QN.

Savage, D. (1999) 'Too chicken to come out?', *Out*, 63, February, p. 34.

Savran, D. (1996) 'The sadomasochist in the closet: white masculinity and the culture of victimization', *Differences: A Journal of Feminist Cultural Studies*, 8, pp. 127–52.

Schwartz, B. (1994) 'Where is cultural studies?', *Cultural Studies*, 8, pp. 377–93.

Schwartz, R. (1996) 'Body, space, and idea', in R. Allsopp and S. deLahunta (eds) *The Connected Body?* (Amsterdam: Amsterdam School of the Arts), pp. 77–81.

Scott, J. (1990) 'The evidence of experience', in H. Abelove, M. Barale and D. Halperin (eds) *The Lesbian and Gay Studies Reader* (London: Routledge), pp. 397–415.

Searle, J. (1969) *Speech Acts: An Essay in the Philosophy of Language* (Cambridge: Cambridge University Press).

—— (1979) 'Metaphor', in A. Ortnoy (ed.) *Metaphor and Thought* (Cambridge: Cambridge University Press).

Sebeock, T. (1986) *The Encyclopedic Dictionary of Semiotics* (Berlin: Monton de Gruyer).

Sedgwick, E. (1990) *Epistemology of the Closet* (Berkeley: University of California Press).

—— (1993) 'Queer performativity: Henry James' *The Art of The Novel*', *GLQ: A Journal of Lesbian and Gay Studies*, 1, pp. 1–16.

—— (1994) *Tendencies* (London: Routledge).

Selden, R. and Widdowson, P. (1993) *A Reader's Guide to Contemporary Literary Theory*, 3rd edition (Lexington: University of Kentucky Press).

Shibles, W. A. (1971a) *An Analysis of Metaphor in the Light of W. M. Urban's Theories* (The Hague: Mouton Press).

—— (1971b) *Metaphor: An Annotated Bibliography and History* (Whitewater, WI: Language Press).

Siegel, L. (1998) 'The gay science: queer theory, literature, and the sexualization of everything', *The New Republic*, 219, no. 19, issue 4,373, 9 November, pp. 30–41.

Signorile, M. (1993) *Queer in America: Sex, the Media, and the Closets of Power* (New York: Random House).

—— (1995) *Outing Yourself* (New York: Fireside, Simon and Schuster).

Silber, I. F. (1995) 'Spaces, fields, boundaries: the rise of spatial metaphors in contemporary sociological theory', *Social Research*, 62, pp. 323–55.

Sim, S. (ed.) (1998) *The Icon Critical Dictionary of Postmodern Thought* (Cambridge: Icon Books).

Sinfield, A. (1989) 'Thatcher, culture, and the closet', *New Statesman and Society*, 2, no. 71, pp. 29–31.

Smith, J. (1996) 'Geographical rhetoric: modes and tropes of appeal', *Annals of the Association of American Geographers*, 86, pp. 1–20.

Smith, N. (1996) *The New Urban Frontier: Gentrification and the Revanchist City* (London: Routledge).

—— (1997) 'Book review of Katherine Kirby's *Indifferent Boundaries*,' *Gender, Place, and Culture*, 4, pp. 253–5.

Smith, N. and Katz, C. (1993) 'Grounding metaphor: towards a spatialized politics', in M. Keith and S. Pile (eds) *Place and the Politics of Identity* (London: Routledge), pp. 67–83.

Soja, E. (1996) *Thirdspace* (Oxford: Blackwell).

Solomon, J. (1989) 'Staking ground: the politics of space in Virginia Woolf's *A Room of One's Own* and *Three Guineas*', *Women's Studies*, 16, pp. 331–47.

Sontag, S. (1990) *Illness as Metaphor* (New York: Doubleday).

Stambovsky, P. (1988) *The Deceptive Image: Metaphor and Literary Experience* (Amherst: University of Massachusetts Press).

States, B. (1996) 'Performance as metaphor', *Theatre Journal*, 48, pp. 1–26.

Stone, S. (1991) 'The empire strikes back: a posttranssexual manifesto', in J. Epstein and K. Straub (eds) *Body Guards: The Cultural Politics of Gender Ambiguity* (London: Routledge), pp. 280–304.

Swanson, D. (1978) 'Toward a psychology of metaphor', in S. Sacks (ed.) *On Metaphor* (Chicago: University of Chicago Press), pp. 161–4.

Tang, W. (1997) 'The Foucauldian concept of governmentality and spatial practices: an introductory note', Occasional Paper, Department of Geography, The Chinese University of Hong Kong.

Tatchell, P. (1991) 'Equal rights for all: strategies for lesbian and gay equality in Britain', in K. Plummer (ed.) *Modern Homosexualities* (London: Routledge), pp. 237–45.

Thompson, F. (1993) 'Metaphors of space: polarization, dualism, and Third World cinema', *Screen*, 34, pp. 38–53.

Tolaas, J. (1991) 'Notes on the origin of some spatialization metaphors', *Metaphor and Symbolic Activity*, 6, pp. 203–18.

Travis, A. (1998) 'Public backs gays in Cabinet', *The Guardian*, 10 November, p. 1.

Trumbo, J. (1998) 'Describing multimedia: the use of spatial metaphors and multimedia design', *News Photographer*, 53, pp. 7–10.

Turbayne, C. M. (1962) *The Myth of Metaphor* (New Haven: Yale University Press).

US census (1990) *Census of Population and Housing, 1990 CPH 3-301A* (Washington, DC: United States Department of Commerce).

Valentine, G. (1996) '(Re)negotiating the heterosexual street', in N. Duncan (ed.) *BodySpace* (London: Routledge), pp. 146–55.

VanDecimeter, J. (1999) 'Wait a minute, Mr. Post-Gay man', *Suck*, Monday, 28 June. Available online at: wiredmail-info@lists.wired.com, pp. 1–11.

Van Gelder, L. (1991) *Are You Two Together?* (New York: Random House).

Vervaeke, J. and Kennedy, J. (1996) 'Metaphors in language and thought: falsification and multiple meanings', *Metaphor and Symbolic Activity*, 11, pp. 273–84.

Waggoner, J. (1990) 'Interaction theories of metaphor: psychological perspectives', *Metaphor and Symbolic Activity*, 5, pp. 91–108.

Wallace, D. (1988) 'Secret gardens and other symbols of gender in literature', *Metaphor and Symbolic Activity*, 3, pp. 135–45.

Wallis, M. (1989) 'Gramsci-the-goalie: reflections in the bath on gays, the Labour party, and socialism', in S. Shepherd and M. Wallis (eds) *Coming on Strong: Gay Politics and Culture* (Boston: Unwin, Hyman), pp. 287–300.

Warner, M. (1994) 'Introduction', in M. Warner (ed.) *Fear of a Queer Planet* (Minneapolis: University of Minnesota Press), pp. vii–xxxi.

Weeks, J. (1978) 'Preface', in G. Hocquenghem, *Homosexual Desire*, originally published 1972 (Chapel Hill: Duke University Press), pp. 23–47.

Weston, K. (1991) *Families We Choose* (New York: Columbia University Press).

Wheelwright, P. (1960) 'Semantics and ontology', in C. Knights and B. Cottle (eds) *Metaphor and Symbol* (London: Butterworths), pp. 1–9.

White, R. (1996) *The Structure of Metaphor* (Oxford: Blackwell).

Wielgosz, A. (1996) 'The topography of writing: Raymond Federman's *The Voice in the Closet*', *Critique*, 37, pp. 108–13.

Willbern, D. (1989) 'Reading after Freud', in G. D. Atkins and L. Morrow (eds) *Contemporary Literary Theory* (Amherst: University of Massachusetts Press), pp. 158–79.

Wincapaw, C. (forthcoming) 'Lesbian, bisexual and women's electronic mailing lists as sexualized spaces', *Journal of Lesbian Studies*.

Wolch, J. and Dear, M. (1988) *The Power of Geography: How Territory Shapes Social Life* (Boston: Unwin Hyman).

Worf, B. (1956) *Language, Thought and Reality* (Cambridge, MA: MIT Press).

Wright, E. (1984) *Psychoanalytic Criticism: Theory and Practice* (London: Routledge).

Zukin, S. (1996) *The Culture of Cities* (Oxford: Blackwell).

# INDEX

Printed and bound by CPI Group (UK) Ltd, Croydon, CR0 4YY

01/11/2024

01782635-0018